Computational Intelligence and Smart Technologies in Solar Thermal Systems

The text covers emerging technologies and innovations in the field, such as hybrid solar thermal systems, advanced materials for collectors, novel heat storage solutions, and advancements in concentrating solar power (CSP) technologies. It presents artificial intelligence and machine learning techniques for solar thermal system optimization.

Key Features:

- Discusses the use of smart technologies such as artificial intelligence and machine learning in transforming solar thermal systems by enhancing their efficiency, reliability, and cost-effectiveness
- Provides insights into optimization techniques for online testing of parameters for solar dryers, AI-based predictive modeling for accurate solar power generation, dual-axis solar panel tracking and monitoring systems, IoT in solar dryer design, and greenhouse-type solar dryer systems
- Covers state-of-the-art computational models used in optimizing solar thermal systems, such as neural networks, genetic algorithms for predictive modeling, control strategies, and system optimization
- Presents case studies and examples of implementations of computational intelligence and smart technologies in solar thermal applications, having real-time applications in a 4kW hybrid solar EV charging station, of five-level inverter for solar photovoltaic systems, agrivoltaic systems, and wind-solar-hydrogen hybrid energy generation
- Introduces simulation and modeling tools used in the design, analysis, and performance evaluation of solar thermal collectors for decentralized cooking, next-generation solar power plants, and hybrid energy generation systems

This book is primarily written for senior undergraduates, graduate students, and academic researchers in mechanical engineering, energy engineering, computer science and engineering, renewable energy, solar energy, and clean technologies.

Solar Thermal Energy Systems: Advancements in Engineering, Ergonomics, and Sustainable Development

Varun Pratap Singh and Ashwani Kumar

The series is a comprehensive exploration encompassing the entire spectrum of solar thermal energy, aiming to bridge traditional fundamentals with modern innovations. It will discuss from the foundational principles of solar radiation and heat transfer mechanisms to the latest advancements in collector technologies, energy storage systems, and integration with hybrid setups. The series will delve into computational modeling, diverse applications in heating, power generation, and economic viability, while also exploring emerging trends such as artificial intelligence integration. Aimed at senior undergraduate students, graduate students, academic researchers and professionals, the proposed series will cover important topics including digital thermal storage innovations, energy policy, artificial intelligence in power station, energy efficient technologies, and solar energy.

Computational Intelligence and Smart Technologies in Solar Thermal Systems: Principles and Applications
Bharti Sharma, Utku Kose, Varun Pratap Singh, and Ashwani Kumar

For more information about this series, please visit: https://www.routledge.com/solar-thermal-energy-systems-advancements-in-engineering-ergonomics-and-sustianable-development/book-series/%20STES

Computational Intelligence and Smart Technologies in Solar Thermal Systems

Principles and Applications

Edited by

Bharti Sharma, Utku Kose,
Varun Pratap Singh, and Ashwani Kumar

CRC Press
Taylor & Francis Group
Boca Raton London New York

CRC Press is an imprint of the
Taylor & Francis Group, an **informa** business

First edition published 2026
by CRC Press
2385 NW Executive Center Drive, Suite 320, Boca Raton FL 33431

and by CRC Press
4 Park Square, Milton Park, Abingdon, Oxon, OX14 4RN

CRC Press is an imprint of Taylor & Francis Group, LLC

ISBN: 9781041062899 (hbk)
ISBN: 9781041062905 (pbk)
ISBN: 9781003634737 (ebk)

DOI: 10.1201/9781003634737

Typeset in Sabon
by Apex CoVantage, LLC

Contents

Aim and Scope

The primary aim of *Computational Intelligence and Smart Technologies in Solar Thermal Systems: Principles and Applications* is to bridge the knowledge gap between the fields of computational intelligence, specifically artificial intelligence (AI) and machine learning (ML), and solar thermal energy systems. This book endeavors to elucidate how AI and ML technologies can be harnessed to enhance the efficiency, reliability, and sustainability of solar thermal energy systems. It aims to provide a comprehensive guide that not only introduces the foundational principles of solar thermal energy and computational intelligence but also explores the cutting-edge methodologies and applications that AI and ML offer to this green technology. Through this exploration, the book seeks to empower engineers, researchers, and practitioners with the knowledge and tools to innovate and implement smart solar thermal systems that contribute to the global shift toward sustainable energy solutions.

The scope of the proposed book encompasses a comprehensive examination of the intersection between AI, ML, and solar thermal energy systems. It explores the fundamentals of both solar thermal technology and computational intelligence, exploring how AI and ML methods can optimize system efficiency, reliability, and sustainability. The book covers a wide range of topics, including optimization techniques for system performance, predictive maintenance, integration with smart grids, and the challenges and opportunities in leveraging AI and ML for solar thermal energy. It also provides insights into real-world applications, showcasing practical implementations and offering guidelines for practitioners. By presenting emerging trends and future directions, the book aims to foster innovation and support the global shift toward smart, sustainable energy solutions, making it a valuable resource for engineers, researchers, and industry professionals involved in renewable energy technologies.

The book is structured into 14 distinct chapters, each dedicated to a fundamental aspect of this interdisciplinary convergence. Chapter 1 of the book introduces readers to the essential principles underpinning solar thermal energy, including its operational mechanisms, types of solar thermal systems, and diverse applications. Simultaneously, it provides a foundational overview of AI and ML, establishing a baseline understanding of these computational technologies. This chapter is designed to equip readers with the necessary background to navigate the complex landscape of solar thermal optimization and innovation through computational intelligence.

Diving into the core subject matter, the second part, Chapters 2–8, extensively covers the application of various AI and ML methodologies to solar thermal systems. Topics include machine learning and IoT in solar dryer systems, computer-based interface systems for online testing of parameters, artificial intelligence-based predictive modeling, optimization of solar energy using a dual-axis solar panel tracking and monitoring system, optimized design and modeling of a 4kW hybrid solar EV charging station, and performance assessment of five-level inverter for solar photovoltaic systems. These chapters emphasize the optimization of solar thermal energy capture, storage, and conversion processes, showcasing how computational intelligence can significantly enhance system performance and sustainability.

The final part, Chapters 9–14, addresses the broader context of solar thermal systems within the renewable energy ecosystem, particularly their integration with smart grids and hybrid energy systems. This includes an in-depth discussion on environmental performance evaluation of agrivoltaic systems for hot arid regions, next-generation solar power plants, hybrid materials for next-generation sustainable solar cells, solar thermal collectors for off-grid applications, advances in solar dryer design and technologies, and wind-solar-hydrogen hybrid energy generation systems.

The book concludes by looking forward to emerging trends, highlighting ongoing research opportunities, and contemplating the future impact of computational intelligence on the evolution of solar thermal energy technologies. Through this structured approach, *Computational Intelligence and Smart Technologies in Solar Thermal Systems: Principles and Applications* aims to provide a comprehensive resource for professionals, researchers, and students in the fields of renewable energy, thermal energy storage, and computational technology. By offering both introductory knowledge and advanced technical insights, the book aspires to foster innovation and support the development of more efficient, reliable, and sustainable solar thermal energy solutions.

Preface

In the quest for sustainable energy solutions, solar thermal systems have emerged as a key technology, harnessing the sun's power to provide heating, cooling, and electricity across various applications. However, the efficiency and optimization of these systems remain perennial challenges, necessitating innovative approaches to enhance their performance and integration into the broader energy ecosystem. It is within this context that *Computational Intelligence and Smart Technologies in Solar Thermal Systems: Principles and Applications* was conceived.

The genesis of this book lies in the recognition of the transformative potential of artificial intelligence (AI) and machine learning (ML) in revolutionizing solar thermal technologies. As computational intelligence continues to evolve, its application in solar thermal systems offers unprecedented opportunities for optimization, predictive maintenance, and smart integration with global energy networks. This book aims to bridge the gap between the complex theoretical underpinnings of AI and ML and their practical applications in improving solar thermal energy systems. Our journey in compiling this work involved extensive research, collaboration with experts in the fields of renewable energy, AI, and ML, and a deep dive into both the technical and practical aspects of solar thermal systems.

This book covers a broad spectrum of topics, from the basics of solar thermal energy and computational intelligence to sophisticated AI and ML algorithms tailored for system optimization and reliability within 14 chapters. Chapters 1 to 8 present the foundation stone of the book, showcasing the integration of AI and ML with solar thermal energy systems. Chapter 1 introduces the burgeoning field of smart solar thermal energy technologies. It explores diverse applications, from heating and power generation to industrial processes. The chapter emphasizes the integration of AI and IoT for enhanced efficiency and reliability. Future directions focus on advanced materials, intelligent control systems, and sustainable energy solutions, highlighting the transformative potential of these technologies. Chapter 2 reviews machine learning's role in greenhouse-type solar dryers. It explores how ML optimizes drying processes, enhancing efficiency and product quality. The chapter details various ML models used for predicting drying kinetics, temperature, and humidity, offering insights into their practical applications and potential for advancing sustainable drying technologies. Chapter 3 explores how sensor data, remote monitoring, and automated control enhance drying efficiency and product quality. Real-time data on temperature, humidity, and airflow enables precise adjustments. This chapter highlights IoT's potential to optimize solar drying processes, reducing energy consumption and ensuring consistent, high-quality dried products. Chapter 4 explores the development of a computer-based interface for online monitoring of solar dryer parameters. This system enables real-time data acquisition and analysis, crucial for optimizing drying processes. It details the hardware and software components, focusing on sensor integration and user-friendly interface design. The aim is to enhance efficiency and control in solar drying applications through precise, automated parameter tracking. Chapter 5 explores AI-driven solar power prediction, comparing support vector regression (SVR), artificial neural networks (ANN), and Random Forest Regression (RFR). It explores their methodologies, highlighting each model's strengths and weaknesses in forecasting solar generation. Chapter 6 explores maximizing solar energy capture through a dual-axis tracking system. It explores optimization strategies, detailing the system's design and functionality. Real-time monitoring enhances performance, ensuring optimal panel alignment with the sun. This chapter examines the system's efficiency gains and practical applications, highlighting its role in advancing solar energy utilization. Chapter 7 details the optimized design and modeling of a 4kW hybrid solar EV charging station. It explores system architecture, component selection, and performance analysis. Utilizing simulation tools, it optimizes energy flow and efficiency. The chapter showcases the integration of solar PV and grid power, demonstrating a sustainable and reliable EV charging solution. Chapter 8 highlights the performance of a five-level inverter in solar photovoltaic systems. It analyzes the inverter's efficiency, harmonic distortion, and switching behavior under varying solar irradiance and load conditions. The chapter presents experimental results and simulations, demonstrating the inverter's suitability for grid-connected and standalone PV applications, highlighting its advantages in improving power quality and reducing losses.

Chapter 9 explores the environmental impact of agrivoltaic systems in hot, arid climates. It evaluates performance through key metrics, analyzing how integrating solar panels with agriculture affects resource use, land efficiency, and ecosystem health. The chapter explores the balance between energy production and agricultural yield, assessing the sustainability of this dual-purpose land use in challenging environments. Chapter 10 explores the frontier of solar energy, exploring next-generation power plants. It examines technological advancements like tandem cells and AI-driven optimization, alongside challenges such as grid integration and material sustainability.

Future aspects, including space-based solar and advanced storage, are discussed, outlining the path toward a more efficient and sustainable solar-powered future. In continuation, Chapter 11 explores the cutting-edge of solar cell technology, focusing on novel hybrid materials. It delves into the design and application of these materials, aiming to enhance efficiency and sustainability in next-generation solar cells. The chapter examines advancements in material science, highlighting their potential to revolutionize solar energy conversion and contribute to a greener future. Chapter 12 explores solar thermal collectors for decentralized cooking, a critical need in many regions. It reviews diverse designs, from simple box cookers to advanced concentrating systems, analyzing their efficiency and suitability. Implementation strategies, including material choices, local manufacturing, and user adoption, are discussed. This chapter highlights the potential of solar cooking to address energy poverty and promote sustainable living. Chapter 13 offers a comprehensive review of solar dryer advancements. It examines diverse designs, from traditional to cutting-edge technologies, emphasizing efficiency and application. This chapter analyzes performance metrics, material innovations, and integrated systems, highlighting the role of solar drying in sustainable agriculture and food preservation. It provides a critical overview of current trends and future directions in solar dryer technology. The conclusive Chapter 14 explores the promising future of wind-solar-hydrogen hybrid energy systems. It analyzes the integration of these renewable sources for stable, continuous power. The chapter delves into system design, efficiency, and economic viability, highlighting hydrogen's role in energy storage. It assesses current technologies and future prospects, emphasizing the potential for sustainable, off-grid solutions and grid stabilization, paving the way for a cleaner energy landscape.

The field of solar thermal energy is at an inflection point, with AI and ML poised to propel it into a new era of efficiency and sustainability. We hope that this book will catalyze further innovation and exploration, contributing to the global transition toward a more sustainable energy future.

About the Editors

Bharti Sharma earned a Ph.D. in Computer Science and Engineering from Uttarakhand Technical University, Dehradun. She is serving in DIT University, Dehradun, as Associate Professor and Academic Head of the CA department. She has 21 years of rich experience in academics and research. Her research work is mainly associated with machine learning (ML) and AI applications, natural language processing (NLP), intelligent transportation systems (ITS), big data analytics, and the respective application domains. In her research work, her team is mainly engaged in the development of intelligent computational models using machine learning methods to solve various challenging problems related to transportation and other relevant areas. She has credited her authorship in various quality publications in international journals, conferences, and book chapters. Presently, she is also working as Editor for many research books. She has also served as Keynote Speaker and Technical and Program Chair Committee Member in many international conferences. Also, she is an active reviewer in various reputed international journals and conferences. She also has rich experience in conducting the Faculty Development Program (FDP) and several workshops at the national level.

Utku Kose, an esteemed Turkish academic born on March 26, 1985, is renowned for his profound contributions to the field of Computer Engineering. Holding the position of Full Professor at Süleyman Demirel University's Faculty of Engineering, Department of Computer Engineering, he also served as a Visiting Researcher at the University of North Dakota, United States. His academic journey began with a Bachelor's degree from Gazi University, followed by a Master's from Afyon Kocatepe University, culminating in a Ph.D. from Selçuk University, all in Turkey. His research, deeply rooted in artificial intelligence, spans across various disciplines, including biomedical AI, machine ethics, and intelligent systems, reflecting his passion for exploring the intersections of technology with ethical considerations and existential risks. Dr. Kose has not only made significant strides in academia through his teaching and research but also holds key editorial and leadership roles in esteemed journals like the *International Journal of Information Communication Technology and Digital Convergence* and the *Journal of Multidisciplinary Developments*. His commitment to advancing knowledge is further demonstrated by his role as a book series editor on topics such as Computational Modeling Applications for Existential Risks and Biomedical and Robotics Healthcare. In addition to his scholarly pursuits, Dr. Kose has contributed significantly to the field through a variety of sectoral experiences, showcasing his versatility and commitment to applying theoretical knowledge in practical settings. His contributions have been recognized globally, further evidenced by his memberships in prestigious organizations such as IEEE and ACM, underscoring his status as a leader in the field. Dr. Kose's research interests are broad yet deeply interconnected, focusing on the safe and ethical advancement of artificial intelligence, the development of intelligent optimization algorithms, and the exploration of AI's philosophical implications. His work is not only theoretical but also practical, as seen in projects like "Artificial Intelligence Based Digital Twin for Healthcare" and the development of "Trustworthy Artificial Intelligence Interaction in Biomedical" systems, highlighting his dedication to leveraging AI for the betterment of society. Through his extensive publications in high-impact journals and conferences, Dr. Kose has established himself as a pioneering figure in his field, dedicated to advancing our understanding and application of AI in ways that are ethically responsible and beneficial to humanity.

Varun Pratap Singh is an assistant professor with the Mechanical Engineering cluster at the University of Petroleum and Energy Studies (UPES), Dehradun, India, and also associated as a PDF with the Solar Thermal Energy Research Group at the University of Stellenbosch, South Africa, renowned as one of the top energy research centres in the world, specializing in solar thermal energy research. Dr. Singh started working on solar thermal technology as his major project with his bachelor's (B.Tech. Mechanical Engineering, Rank Holder) and continued his passion with a master's (M.Tech. Thermal, Mechanical Engineering, Gold Medallist), and during his Ph.D. (Mechanical Engineering-Solar Thermal) by conducting thermohydraulic analysis of different roughnesses in solar air heaters, respectively. His experience includes 13 years of teaching, seven years of research and industrial projects, and seven years of administrative experience as the centre head of a Center of Excellence (CoE), industry-academia coordinator, and responsible for industry-academia integration at various fronts.

He has published 30 research publications (including 16 SCIE and 12 Scopus papers), and five patents, edited three books, and authored more than 15 book chapters for international publishers of repute.

He is a series editor for a book series titled "Solar Thermal Energy Systems: Advancements in Engineering, Ergonomics, and Sustainable Development" with Taylor & Francis. He also served as an editor for the *IJERA* journal, edited two special issues of the SCIE journal *Energies*, and reviewed more than ten SCIE journals. He completed nine projects as principal investigator (PI) and two projects as co-PI for various government research organizations and industry partners with a total research grant of more than 30 lakh INR. He has delivered many invited talks in international and national webinars, FDPs, and workshops and attended 20+ MOOCs and 15+ FDPs from various international and national bodies like DTU-Netherlands, Tsinghua University-China, National Chiao Tung University-Taiwan, IMF, IITs, and AICTE. He has been awarded an award for outstanding mentoring of students and a perfect faculty award for excellence in academics and research. He has successfully guided 23 B.Tech, M.Tech, and Ph.D. projects and theses. His major research interests include the design, development, experimental, and simulation performance assessment of solar thermal systems. His extended research interests include working on climate change and mitigation, carbon capture technology, hydrogen production, LCA and techno-economic analysis, and small and micro-hydro applications with the development of special-purpose machines (SPM). With a substantial background in solar thermal energy and a distinguished affiliation with UPES, India, and Stellenbosch, South Africa, Dr. Singh brings a wealth of knowledge and experience to the table.

Ashwani Kumar holds a Ph.D. in Mechanical Engineering, specializing in Mechanical Vibration and Design, with a diverse and impressive career spanning over 14 years. He currently holds the position of Professor and Head of Department, Mechanical Engineering (Gazetted Officer Class 1), at the Technical Education Department, Uttar Pradesh (under the Government of Uttar Pradesh) in Kanpur, India. Prior to his current role, Dr. Kumar served as Senior Lecturer and Lecturer in the same organization and Assistant Professor in the Department of Mechanical Engineering at Graphic Era University, Dehradun (ranked 52nd by NIRF India), from July 2010 to November 2013. Beyond his academic contributions, Dr. Kumar possesses extensive leadership and administrative experience. He has held numerous key positions within the education sector, including Coordinator for AICTE-Extension of Approval, Nodal Officer for PMKVY-TI Scheme (Government of India), Internal Coordinator—CDTP Scheme (Government of Uttar Pradesh), Industry Academia Relation Officer, Assistant Centre Superintendent (ACS)—Institute Examination Cell, Zonal Officer for Joint Entrance Examination (JEE-Diploma), and Sector Magistrate for State Assembly and General Elections. Dr. Kumar is a passionate researcher and a prominent figure in the field of academic publishing. His dedication to knowledge dissemination is evident in his role as Series Editor for eight distinguished book series published by reputed international publishers. These series encompass a wide range of topics, titled: Advances in Manufacturing, Design and Computational Intelligence Techniques; Renewable and Sustainable Energy Developments; Smart Innovations and Technological Advancements in Mechanical and Materials Engineering; Solar Thermal Energy Systems: Advancements in Engineering, Ergonomics, and Sustainable Development; Artificial Intelligence and Machine Learning for Intelligent Engineering Systems; Computational Intelligence and Biomedical Engineering; Advances in Green Technologies for Sustainable Energy Solutions; and Advances in Materials, Manufacturing and Computational Intelligence Techniques. Dr. Kumar's editorial expertise extends beyond book series. He serves as Editor-in-Chief for the *International Journal of Materials, Manufacturing and Sustainable Technologies* (IJMMST, *ISSN: 2583–6625*), and Editor of the *International Journal of Energy Resources Applications* (IJERA, *ISSN: 2583–6617*). He has also been a guest editor for a special issue of *Buildings* titled "Sustainable Buildings, Resilient Cities and Infrastructure Systems" (*ISSN: 2075–5309*, I.F. 3.1) and holds editorial board positions on eight international journals and acts as review board member of 20 prestigious (Indexed in SCI/SCIE/Scopus) international journals, with high impact factor. Dr. Kumar's dedication to scholarship is further demonstrated by his prolific authorship of over 200+ articles in prestigious journals, book chapters, and conference proceedings. Additionally, he has co-authored or edited over 40+ books in Mechanical Engineering, Materials Science, and Renewable Energy Engineering. Dr. Kumar's achievements have been recognized with prestigious awards, including a Best Teacher award for excellence in academics and research. He has also successfully guided 15 students through their B.Tech, M.Tech, and Ph.D. theses and currently serves as an external doctoral committee member at S.R.M. University, New Delhi. Dr. Kumar is actively involved in cutting-edge research, focusing on areas such as artificial intelligence and machine learning in mechanical engineering, smart materials and manufacturing techniques, thermal energy storage, building efficiency, renewable energy harvesting, sustainable transportation, and heavy vehicle dynamics. He holds four patents and frequently participates in international conferences as an invited speaker, session chair, and member of advisory boards, review boards, and program committees.

Contributors

Saquibahmed Bagali, Department of Electrical & Electronics Engineering, Jain College of Engineering, Belagavi, Karnataka, India

Prashant Baredar, Professor & Head, Energy Centre, Maulana Azad National Institute of Technology, Bhopal, Madhya Pradesh, India

V. N. Bartaria, Professor, Mechanical Engineering, LNCT University, Bhopal, Madhya Pradesh, India

Dr. Uppin Chandrashekhar, Associate Professor, Hensard University, Nigeria

Anurag Chauhan, Department of Electrical Engineering, Rajkiya Engineering College, Banda, Uttar Pradesh, India

Aviral Chaurasia, Centre for Advanced Studies, Lucknow, India

Venkatratnam Chitturi, Department of Electrical & Electronics Engineering, Jain College of Engineering, Belagavi, Karnataka, India

Gaurav Dwivedi, Assistant Professor, Energy Centre, Maulana Azad National Institute of Technology, Bhopal, Madhya Pradesh, India

Akshat Garg, Department of Mechanical Engineering, Amity University, Uttar Pradesh, Noida, India

Saksham Gaur, Amity University, Uttar Pradesh, Noida, India

Yatika Gori, Assistant Professor, Department of Mechanical Engineering, Graphic Era Deemed to Be University, Dehradun, Uttarakhand, India

Albar Khan, Department of Mechanical Engineering, Amity University, Uttar Pradesh, Noida, India

Ashwani Kumar, Professor and Head of Department, Mechanical Engineering, Technical Education Department, Uttar Pradesh, Kanpur, India

Shan Mahto, Energy Centre, Maulana Azad National Institute of Technology, Bhopal, Madhya Pradesh, India

Saurabh Mishra, Centre for Advanced Studies, Lucknow, India

Shubham Mishra, Research Scholar, Department of Mechanical Engineering, LNCT University, Bhopal, India

Prof. Dileep Kumar Mohanachandran, Professor Management and Technology, Hensard University, Sagbama L.G.A., Bayelsa State, Nigeria, International Research Fellow, SEGi University, Malaysia

Achyutha Prasad N, Professor, Department of Computer Science and Engineering, East West Institute of Technology, Bangalore, India

Sachin Kumar Nikam, Assistant Professor, Department of Mechanical Engineering, LNCT University, Bhopal, India

Rishu Pandey, Department of Electrical Engineering, JSS Academy of Technical Education, Noida, Uttar Pradesh, India

Abhilasha Pawar, Department of Electrical Engineering, JSS Academy of Technical Education, Noida, Uttar Pradesh, India

Surrendra Poonia, Central Arid Zone and Research Institute, Jodhpur, Rajasthan, India

Neha Singh Raghuvanshi, ABES Engineering College, Ghaziabad, Uttar Pradesh, India

Satish Pal Singh Rajput, Dean, Institutional Development & International Relations, Maulana Azad National Institute of Technology, Bhopal, Madhya Pradesh, India

Amit Kumar Roy, Department of Electrical Engineering, JSS Academy of Technical Education, Noida, Uttar Pradesh, India

Shresth Saini, Amity University, Uttar Pradesh, Noida, India

Sakshi Sarathe, Energy Centre, Maulana Azad National Institute of Technology, Bhopal, Madhya Pradesh, India

Bharti Sharma, Associate Professor, Department of Mechanical Engineering, DIT University, Dehradun, Uttarakhand, India

Meeta Sharma, Amity University, Uttar Pradesh, Noida, India

Anoop Kumar Shukla, Department of Mechanical Engineering, Amity University, Uttar Pradesh, Noida, India

A.K. Singh, Central Arid Zone and Research Institute, Jodhpur, Rajasthan, India

Digvijay Singh, School of Engineering and Technology, K.R. Mangalam University, Gurugram, Haryana, India

Ishita Singh, Department of Electrical Engineering, JSS Academy of Technical Education, Noida, Uttar Pradesh, India

Neetu Singh, Research Scholar, LNCT University, Bhopal, Madhya Pradesh, India

Varun Pratap Singh, Assistant Professor, Department of Mechanical Engineering, University of Petroleum and Energy Studies, Dehradun, Uttarakhand, India & Department of Mechanical & Mechatronics Engineering, Stellenbosch University, South Africa

Ajeet Kumar Soni, Registrar, LNCT University, Bhopal, Madhya Pradesh, India

Ashish Srivastava, Department of Electrical Engineering, Rajkiya Engineering College, Banda, Uttar Pradesh, India

Dr. Gopala T, Assistant Professor, Department of Computer Science and Engineering, Sri Siddhartha Institute of Technology, Tumakuru, Bangalore, India

Ashish Tewari, Department of Electrical Engineering, JSS Academy of Technical Education, Noida, Uttar Pradesh, India

Divyansh Tiwari, Department of Electrical Engineering, JSS Academy of Technical Education, Noida, Uttar Pradesh, India

Anurag Tripathi, Department of Electrical Engineering, Institute of Engineering & Technology, Lucknow, Uttar Pradesh, India

Subho Upadhyay, Department of Electrical Engineering, Dayalbagh Educational Institute (Deemed to be University), Agra, Uttar Pradesh, India

Krishna Varma, Centre for Technology Alternatives for Rural Areas, Indian Institute of Technology, Bombay, Maharashtra, India

Dr. Kalpana Venugopal, Professor and Associate Dean, School of Business, Commerce & Finance, AIMS Institutes,1st Stage, 1st Cross, Peenya Industrial Area, Bangalore

Acknowledgment

The journey culminating in the publication of *Computational Intelligence and Smart Technologies in Solar Thermal Systems: Principles and Applications* has been an intricate tapestry woven with the contributions of numerous dedicated individuals and supportive organizations. To each and every one who lent their expertise, encouragement, and time, we extend our most profound gratitude.

At the heart of this endeavor lie the contributing authors. Their profound knowledge and unwavering commitment to the exploration of artificial intelligence (AI) and machine learning (ML) within the realm of solar thermal energy systems have been the cornerstone of this book. Each chapter reflects their rigorous research, insightful analysis, and innovative approaches, pushing the boundaries of what is possible in this critical field. Without their collective passion and scholarly dedication, this project would have remained a mere concept, an unfulfilled aspiration. Their contributions have not only enriched the academic discourse but also laid a foundation for future advancements in renewable energy technologies.

We are immensely indebted to our esteemed academic and professional colleagues, whose intellectual generosity and critical feedback have been instrumental in shaping the quality and depth of this work. Their willingness to engage in thoughtful discussions, offer constructive criticism, and share their expertise has been invaluable. Throughout the drafting and revision process, their insights have refined our arguments, clarified complex concepts, and ensured the accuracy and relevance of the presented information. We particularly acknowledge the diligent peer reviewers, whose meticulous assessments and insightful suggestions have played a crucial role in elevating the scholarly rigor of each chapter. Their dedication to maintaining high standards of academic excellence has ensured that this book stands as a reliable and authoritative resource for researchers, practitioners, and students alike.

The institutional support we received has been nothing short of transformative. We express our sincere appreciation to our respective institutions for providing the necessary resources, fostering a collaborative environment, and offering unwavering encouragement throughout this ambitious project. The availability of research facilities, access to cutting-edge technologies, and the intellectual stimulation provided by our institutional communities has been essential to our success. We are grateful for the opportunities afforded to us, which have enabled us to delve deeply into the complexities of solar thermal systems and explore the potential of computational intelligence in this domain.

Finally, and perhaps most importantly, we extend our deepest gratitude to our families and friends. Their unwavering support, patience, and understanding have been the bedrock of our resilience throughout this demanding endeavor. The countless hours spent researching, writing, and revising often necessitated sacrifices, and their encouragement sustained us through these challenging times. Their belief in our work, their willingness to listen, and their ability to provide a sense of balance and perspective have been invaluable. We recognize that this book is as much a testament to their support as it is to our efforts.

Computational Intelligence and Smart Technologies in Solar Thermal Systems: Principles and Applications is a product of collective effort, a symphony of contributions from diverse individuals united by a shared passion for advancing renewable energy technologies. We hope that this work serves as a valuable resource for those seeking to understand and apply computational intelligence and smart technologies in solar thermal systems. We aspire to contribute meaningfully to the ongoing dialogue surrounding sustainable energy solutions and to inspire further innovation in this critical field. It is our sincere hope that this book will serve as a catalyst for future research and development, fostering a brighter and more sustainable future for generations to come.

Chapter 1

Smart Solar Thermal Energy Technologies

Applications and Future Directions

Bharti Sharma, Ashwani Kumar, Varun Pratap Singh, Yatika Gori, Sakshi Sarathe, and Gaurav Dwivedi

1.1 INTRODUCTION

1.1.1 Importance of Renewable Energy Sources

Growing energy demand across the world, together with worries regarding climate change and resource depletion, has created an immense movement toward renewable energy resources. Renewable energy, such as solar, wind, hydro, and biomass, provides a green alternative to fossil fuels, with lower greenhouse gas emissions and reduced environmental footprint. Of these, solar energy is one of the most available and accessible renewable energies. The utilization of solar power through photovoltaic and solar thermal technology is an excellent option for responding to the world's increasing demand for energy and for promoting energy security and sustainability [1].

The transition to renewable energy is also prompted by advances in energy policy, innovation in technologies, and reduction of the cost of clean energy production. Governments of most nations have put policies and incentives in place that support the use of renewable energy, such as tax credits, feed-in tariffs, and renewable portfolio standards. These policies facilitate the process of moving toward a more sustainable energy system. Besides, the falling cost of solar energy production has placed it as a sustainable alternative to traditional sources of energy, strengthening its position in the world energy shift [2].

1.1.2 Overview of Solar Thermal Energy

Solar thermal energy is a technology that produces and uses the heat of the sun for applications such as domestic, commercial, and industrial usage. Solar thermal systems differ from photovoltaic (PV) systems, as they aim to collect heat energy, which in turn can be utilized for heating spaces, heating water, and generating power. The fundamental operating mechanism is the absorption of solar radiation on a heat-absorbing surface that transfers the heat to a working fluid. The thermal energy is storable and can be used later, making the system more reliable and efficient in solar power use.

DOI: 10.1201/9781003634737-1

Solar thermal technologies are divided into three broad categories depending on their operating temperature ranges, each with different applications. Low-temperature solar thermal systems, which operate below 100°C, are mainly applied for domestic hot water and space heating purposes. These systems are extensively used in residential and commercial buildings to save energy costs and mitigate greenhouse gas emissions by harnessing renewable solar energy. They often use flat-plate collectors or evacuated tube collectors to collect and transfer solar heat effectively. Medium-temperature solar thermal systems operating between 100°C and 250°C find extensive application in industries, some of which include process heat, solar cooling, and desalination. Food processing, textile, and pharmaceutical industries, among others, use moderate levels of heat, which are easily provided by these systems. Solar cooling systems, including solar heat-driven absorption chillers, reduce the use of electricity for air conditioning, constituting an energy-saving option in areas with high cooling requirements. Solar desalination machines also employ thermal energy to transform seawater into freshwater, constituting a renewable option for tackling water shortages in desert areas.

High-temperature solar thermal systems, running at over 250°C, are mainly employed in concentrated solar power (CSP) power plants, which produce electricity by utilizing solar heat to generate mechanical power through thermodynamic cycles like the Rankine or Brayton cycle. CSP technologies, such as parabolic troughs, solar towers, and Fresnel reflectors, focus sunlight onto a receiver to generate high-temperature heat, which is used to drive steam turbines or gas cycles for electricity generation. These systems provide a renewable source of power for energy generation, part of a cleaner energy mix. As solar thermal technology advances, gains in efficiency and reductions in cost are making these systems increasingly competitive and viable to deploy on a large scale.

There has been progress in solar thermal technology, making it more efficient at collecting and storing heat. Development of solar absorber materials, heat transfer fluids, and PCMs has all contributed to improved overall efficiency in solar thermal systems. Investigations of nanofluids with added nanoparticles to increase thermal conductivity have also contributed to the enhancement of solar thermal collector efficiency.

1.1.3 Comparison with Other Solar Energy Technologies

Solar thermal energy is a technology that harvests and utilizes the heat of the sun for a number of purposes, ranging from residential, commercial, and industrial applications. In contrast to photovoltaic (PV) systems, which convert sunlight directly into electricity, solar thermal systems are aimed at capturing heat energy, which is later utilized for space heating, water heating, and power generation. The fundamental operation is the absorption of solar radiation into a heat-absorbing surface, which distributes the heat to a working fluid. The thermal energy is meant to be stored and reused later, boosting the efficiency and reliability of solar power utilization [3]. Solar thermal technologies are divided into three categories depending on their operating temperature ranges, each finding different applications. Low-temperature solar thermal systems, which work below 100°C, are mostly applied for space heating and domestic water heating purposes. These systems are extensively utilized in residential as well as commercial buildings to save energy expenses and minimize greenhouse gas emissions by using renewable solar energy. They generally utilize flat-plate collectors or evacuated tube collectors to efficiently capture and transfer solar heat [4].

Medium-temperature solar thermal systems operating in the 100°C to 250°C range are used significantly in industrial applications, ranging from process heat to solar cooling and desalination. A large number of industries, including food processing, textiles, and pharmaceuticals, demand moderate levels of heat, which can be well provided by these systems. Solar cooling systems, including solar-powered absorption chillers, minimize the use of electricity for cooling purposes and thus are an energy-saving option in areas with large cooling loads. Solar desalination plants also make use of thermal energy to transform seawater into fresh water, which is a renewable means of counteracting water shortages in desert areas [5].

High-temperature solar thermal systems, which are above 250°C, are essentially applied in concentrated solar power (CSP) plants, which produce electricity by the transformation of solar heat to mechanical power through thermodynamic cycles like the Rankine or Brayton cycle. CSP technologies like parabolic troughs, solar towers, and Fresnel reflectors focus sunlight onto a receiver to generate high-temperature heat, which is employed to drive steam turbines or gas cycles for electricity generation. These systems provide a renewable source for power generation over fossil fuels and contribute to a more sustainable energy mix [4]. As the technology of solar thermal is advanced, increased efficiency and declining costs make these systems increasingly competitive and viable for mass adaptation [6].

Recent solar thermal innovations have enhanced the efficiency of heat collection and storage. New technologies for solar absorbers, heat transfer fluids, and phase-change materials (PCMs) have improved the overall performance of solar thermal devices. Nanofluid studies, in which particles at the nanoscale improve thermal conductivity, have further enhanced solar thermal collector efficiency [6].

Furthermore, the lifecycle environmental footprint of solar thermal is usually lower than that of PV systems. The manufacturing process of PV panels requires energy-intensive processes and the incorporation of rare materials, while solar thermal collectors usually have smaller carbon footprints and longer lifetimes. Thus, solar thermal power is a desirable choice for minimizing environmental effects with energy security [7].

The subsequent parts of this chapter will explore deeper into the basic principles of solar thermal energy, different collector technologies, energy storage technologies, and recent developments determining the future of solar thermal energy conversion.

1.2 FUNDAMENTALS OF SOLAR RADIATION AND HEAT TRANSFER

1.2.1 Solar Spectrum and Energy Potential

The sun radiates a wide band of electromagnetic radiation, mostly ultraviolet (UV), visible, and infrared (IR). Most of the solar energy that enters Earth is from 290 nm to 3,000 nm and about 43% visible, 52% infrared, and 5% ultraviolet. Solar power that comes to the Earth's upper atmosphere is about 1,361 W/m^2 or the solar constant. But with atmospheric scattering, absorption, and reflection, the solar radiation actually incident on the Earth's surface differs from place to place, hour to hour, and depending on the state of the atmosphere [8].

1.2.2 Mechanisms of Heat Transfer: Conduction, Convection, and Radiation

Heat transfer takes place by three major mechanisms: conduction, convection, and radiation.

1. Conduction—This refers to the flow of heat by means of a solid material without actual movement of the material itself. It results from the collision of vibrating molecules and free electrons. The conductive heat transfer rate is described by Fourier's Law:

$$q = k\frac{dT}{dx}$$

   where q is the heat flux (W/m^2), k is the thermal conductivity (W/m·K), and dT/dx is the temperature gradient [9].
2. Convection—It involves heat transfer via the motion of a fluid (liquid or gas). It is either natural convection (created by buoyancy forces due to temperature differences) or forced convection (resulting from an external force such as a fan or pump).

   Newton's Law of Cooling represents convective heat transfer [9]:

$$q = hA(T_S - \infty)$$

   where h is the convective heat transfer coefficient (W/m^2·K), A is the surface area, Ts is the surface temperature, and T∞ is the ambient temperature.

Radiation—This is the conduction of heat in the form of electromagnetic waves, mainly in the infrared range. Radiation does not need a medium to exist, unlike conduction and convection. The Stefan-Boltzmann Law controls radiative heat transfer:

$$q = \sigma \varepsilon AT4$$

where σ is the Stefan-Boltzmann constant ($5.67 \times 10 - 8W/m^2K4$), ε is the emissivity of the surface, A is the surface area, and T is the absolute temperature in Kelvin [9].

1.2.3 Absorption, Reflection, and Transmission of Solar Radiation

When solar radiation comes into contact with a surface, three general processes happen: absorption, reflection, and transmission. Absorption happens when part of the incident radiation is absorbed and transformed into heat, with the absorption coefficient (α) controlling the energy trapped. Black surfaces, for example, are highly absorptive ($\alpha \approx 0.9 - 0.98$), while white or reflective surfaces are less absorptive ($\alpha \approx 0.1 - 0.3$) [1]. Reflection happens when a part of the solar radiation is reflected off the surface, depending on what properties it has. Reflectivity (ρ) is the ratio of the reflected radiation, high in polished metals and low for dark ones that have higher energy absorption. Transmission is the process where transparent and semi-transparent materials allow some of the incident solar radiation to pass through them. Transmissivity (τ) is the measure of this property. For instance, glass allows a lot of visible light to pass through while absorbing or reflecting infrared radiation [10].

The relationship among these properties is given by the energy balance equation:

$$\alpha + \rho + \tau = 1$$

which ensures energy conservation.

1.3 TYPES OF SOLAR THERMAL COLLECTORS

1.3.1 Flat Plate Collectors

Figure 1.1 illustrates a flat-plate collector, which is a form of solar thermal collector employed for uses such as water heating and space heating. It has various components, among which is the glass plate on top, through which sunlight is permitted to enter but with little heat loss. The frame offers structural support, and the bottom and side insulation prevents heat loss, making it more efficient. Within the collector, a black absorber plate absorbs solar

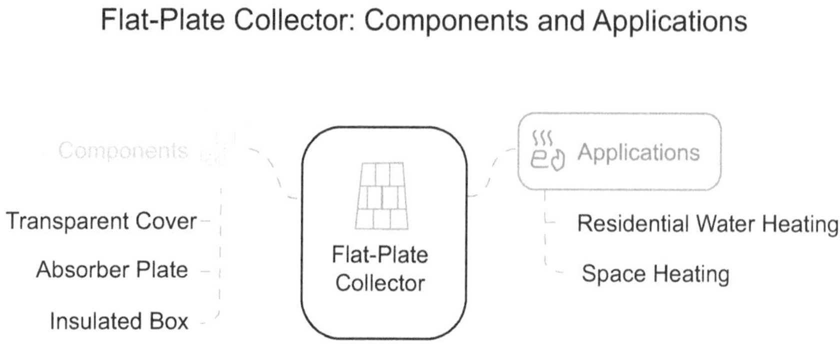

Figure 1.1 Schematic of a flat-plate collector showing key components.

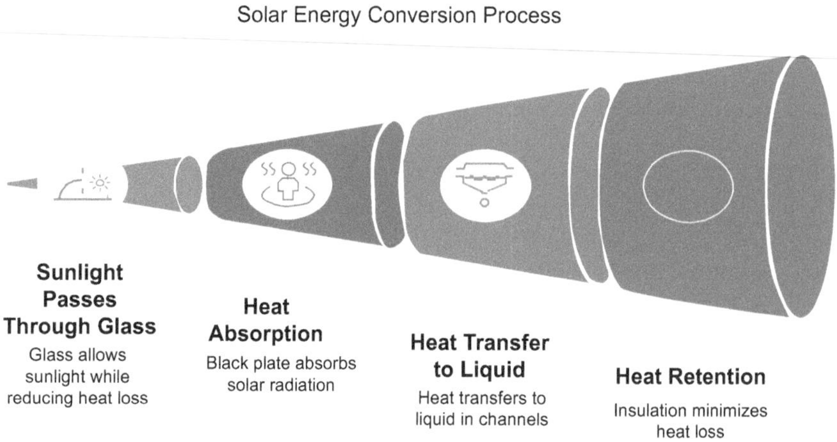

Figure 1.2 Schematic of a flat-plate collector showing key components.

radiation and conveys the heat to a liquid flowing through the heat-conveying liquid channels. The tubes, which are implanted in the absorber plate, transport the heated fluid, which can be utilized for different purposes of heating. The system works by letting the solar energy heat the liquid, which is transported through the channels either by natural circulation or with the help of a pump. The insulation ensures that the heat is retained, and the system becomes more efficient. Flat-plate collectors are best suited for low to moderate temperature applications, usually between 60°C and 100°C [11].

Figure 1.2 depicts the solar energy conversion process, demonstrating how solar energy is harnessed and used for heating. The process begins with sunlight going through glass, where a transparent cover lets the solar radiation in while reducing heat loss. Then, there is absorption of heat as a black plate absorbs the solar radiation, converting it to thermal energy. This energy is then passed on in the heat transfer to the liquid phase, where the heat

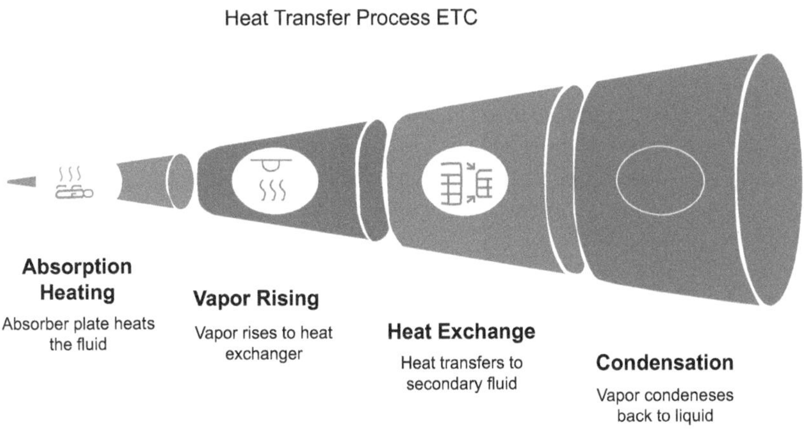

Figure 1.3 Schematic of an evacuated tube collector showing key components and heat transfer process.

absorbed flows into a liquid that moves through channels. Lastly, the system provides heat retention, where insulation reduces heat loss, ensuring efficiency. This process is essential for solar thermal systems, ensuring maximum energy capture and use.

1.3.2 Evacuated Tube Collectors

Figure 1.3 shows the operating principle of an evacuated tube collector (ETC), a very efficient solar thermal collector for heating purposes like water heating and space heating. The system is made up of evacuated glass tubes, each of which has a heat pipe with an absorber plate inside. The evacuated glass tubes produce a vacuum, which serves as an insulator to minimize heat loss and maximize efficiency, even under cold or cloudy conditions. Solar radiation that hits the absorber plate within the evacuated tube warms the fluid inside the heat pipe. This results in the liquid inside the heat pipe evaporating and flowing up toward the heat exchanger (condenser) at the top portion of the collector. Within the heat exchanger, the heated vapor passes thermal energy to an auxiliary heat transfer fluid that's circulated through the network of piping within the enclosed box. The heat is released as the vapor within the heat pipe condenses into a liquid and flows to the lower half of the heat pipe, from where it again gets reheated, repeating the cycle [11].

The system has an inlet where cooler fluid flows into the collector and an outlet through which the hot fluid flows out for storage or used for heating other purposes. Working fluid motion through the system may be naturally induced by convection or supported with a pump based on the design of the system. Evacuated tube collectors are superior to flat-plate collectors in terms of thermal performance since they reduce heat losses and have high temperatures. This renders them particularly beneficial in cold climates where efficiency retention is important. Vacuum insulation surrounding the heat pipes minimizes conductive and convective heat losses substantially, ensuring that a high percentage of the solar energy collected is utilized efficiently for heating purposes [11].

1.3.3 Concentrators

Figure 1.4 shows four kinds of solar concentrating collectors, which are used to concentrate and intensify solar radiation for high-temperature uses like power generation and industrial heating. The first one, indicated in (a) as a parabolic trough collector, is a parabolic trough-shaped reflector that concentrates sunlight on an absorber tube placed at its focal line. The absorber tube holds a heat transfer fluid, which picks up and carries the heat to be utilized in a thermal system. These collectors are utilized on a large scale in solar thermal power plants. The second variety, shown in (b) as a linear Fresnel reflector, uses parallel arrays of flat or slightly curved mirrors to focus sunlight onto a receiver above them. This design is more economical compared to parabolic troughs and is traditionally employed for producing steam or heating fluid. The third one, depicted in (c) as a solar power tower (central receiver system), consists of a number of heliostats (flat mirrors) tracking the sun and reflecting sunlight onto a receiver located at the end of a tall tower. The intense heat is utilized to generate high-temperature steam, which is used to power turbines for electricity generation. Finally, (d) shows a parabolic dish collector, which consists of a parabolic reflector

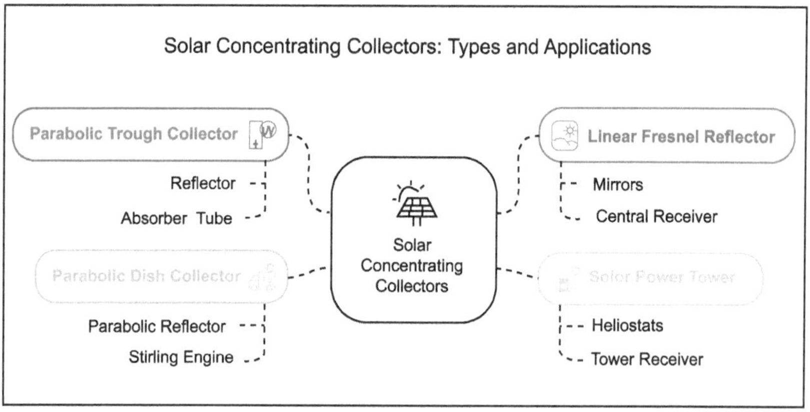

Figure 1.4 Types of solar concentrating collectors—parabolic trough, linear Fresnel, solar tower, and parabolic dish.

that concentrates sunlight onto a small receiver/engine at its focal point. This configuration is commonly employed to power a Stirling engine, and hence it is an effective system for decentralized electricity generation. All these solar concentrators are designed to maximize energy capture by concentrating sunlight onto a reduced receiving area to improve efficiency and allow for higher operating temperatures in different solar thermal applications [11].

1.4 SOLAR THERMAL CONVERSION MECHANISMS

Solar thermal conversion is the process of converting solar energy into thermal energy for different uses. The conversion is done by either direct heating or indirect thermodynamic cycles and is an important aspect of residential, commercial, and industrial applications.

1.4.1 Direct Heating Applications

Direct applications of heating involve solar water heating, solar air heating, and solar space heating. Solar water heaters utilize collectors that absorb the sun's radiation and heat water, which is stored for future usage. Solar water heating systems may be either active, employing pumps, or passive, utilizing natural convection. Research indicates that solar water heating is capable of supplying as much as 80% of the total hot water for a household in areas with high solar radiation. In the same way, solar air heating systems trap solar radiation to warm air, which may be utilized for industrial drying, ventilation, and space heating. Transpired solar collectors and glazed air collectors are popular techniques with a thermal efficiency of 50–70%. Passive and active solar methods such as greenhouse heating, radiant floor heating, and Trombe walls can lower traditional heating energy use by 60% in temperate climates [12].

Figure 1.5 shows several direct solar energy heating applications, which highlight how solar power is used for functional heating in day-to-day applications. Solar energy is captured by the rooftop solar water heating panels to heat water for home and industrial use, lessening the need for electricity or gas-based systems. A solar cooker is shown outside, using the sun's energy to cook food, thus providing an environmentally friendly means compared to conventional fuel-based cooking. Also, the solar drying apparatus is depicted conserving farm produce through the utilization of sunlight to dry crops and fruits, promoting cost-effective and eco-friendly storage of food. The vibrant sunlight and blue sky in the background underscore the widespread availability of solar energy, while the foliage underlines a natural environment. Generally, the picture underscores how direct solar heating uses help promote sustainability, increase energy efficiency, and lower carbon content.

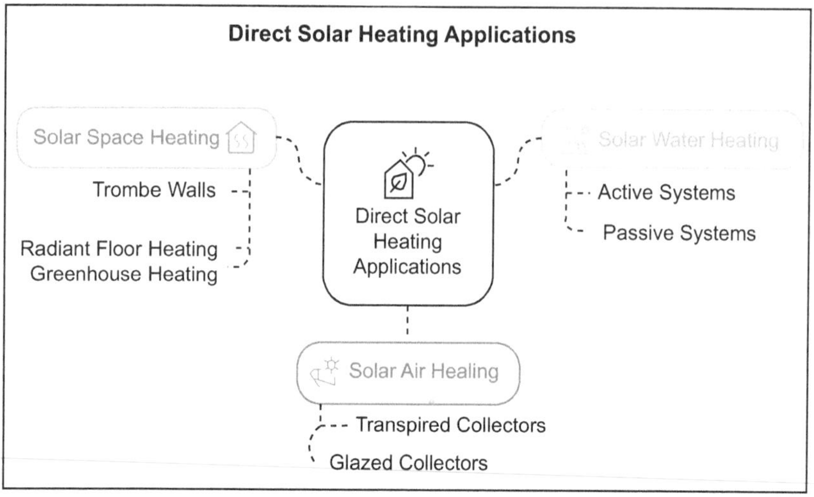

Figure 1.5 Direct heating applications of solar energy.

1.4.2 Indirect Solar Thermal Conversion (Thermodynamic Cycles)

Indirect solar thermal conversion uses thermodynamic cycles like the Rankine and Brayton cycles to produce electricity. For the Rankine cycle, the solar collectors produce heat in a working fluid (e.g., water, molten salt, or organic liquids) to create steam, which powers a turbine. Parabolic trough collectors and linear Fresnel collectors are two popular concentrated solar power (CSP) technologies utilizing this cycle with a thermal efficiency of 30–40%. The Brayton cycle, utilized mainly for solar tower power plants, expands compressed air or a different gas, which is heated and produces electricity. With a higher temperature operation than that of the Rankine cycle, the Brayton cycle can produce over 45% efficiency using high-temperature receiver designs [13].

Figure 1.6 illustrates the Indirect Solar Thermal Conversion Process, outlining how solar energy is converted into electricity in a series of stages. Solar collectors first absorb the sun's energy and transfer the heat to a working fluid. The heated fluid produces steam, which is then utilized to power a turbine. As the expanding steam passes through the turbine, it generates electricity. During the second stage, the residual heat is used in the Brayton cycle, in which the compressed air is heated to further improve energy efficiency. Lastly, the hot air expands through a second turbine, producing more electricity. This process maximizes energy conversion by efficiently utilizing both steam and hot air, thus improving solar thermal systems.

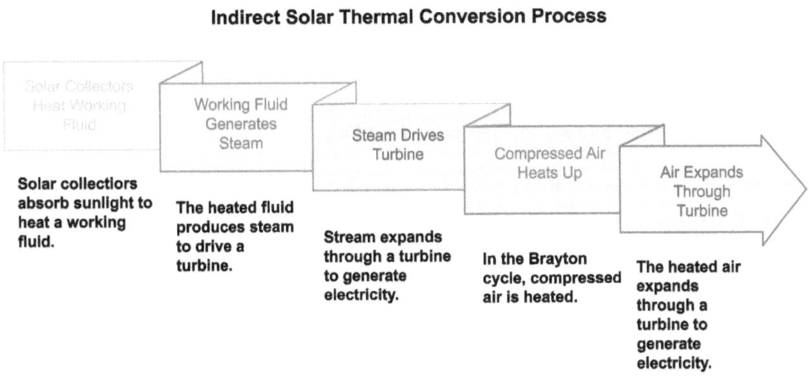

Figure 1.6 Indirect solar thermal conversion through CSP technologies.

1.5 INDUSTRIAL APPLICATIONS OF SOLAR THERMAL ENERGY

Solar thermal energy is finding more use in industrial processes, supplying process heat at multiple temperature levels. Solar energy finds application in the food processing sector for drying, pasteurization, and sterilization, while solar thermal drying systems cut down energy usage by as much as 70% over traditional practices. Solar-heated water and steam are useful in the textile and chemical sectors, enabling processes like dyeing, washing, and chemical reactions. Solar desalination technologies, such as multi-effect distillation (MED) and multi-stage flash (MSF) processes, provide a cost-competitive option compared to fossil-fuel-based desalination, with substantial operation cost reduction. Solar furnaces, which reach temperatures of over 1,000°C, are used in high-temperature industrial processes, including metal processing and cement production. These efficient solar thermal systems assist in decarbonizing industrial heat processes, leading to a 50% reduction in CO_2 emissions [14].

1.6 THERMAL ENERGY STORAGE (TES) SYSTEMS

Thermal Energy Storage (TES) systems play a critical role in the promotion of the efficiency and reliability of renewable energy technologies by accumulating excess thermal energy for future use. TES enables improved energy management for solar thermal power generation, process heating in industry, and space heating and cooling in buildings. The three main types of TES systems are sensible heat storage (SHS), latent heat storage (LHS), and thermochemical energy storage (TCES), and each one has merits in terms of storage capacity, range of temperature, and application [15].

1.6.1 Sensible Heat Storage (SHS)

Sensible heat storage (SHS) is the most common thermal energy storage (TES) technique, in which thermal energy is stored by heating a storage medium without altering its phase. The energy stored is a function of the specific heat capacity, mass, and temperature change of the storage material. Water, molten salts, and rocks are typical SHS materials. Water-based SHS is very efficient as a result of water's good specific heat capacity and abundance and is used in domestic and industrial hot water storage systems, solar water heating, and district heating [16]. Molten salt storage, usually employing a sodium nitrate-knowledge nitrate mixture, is utilized in most concentrated solar power (CSP) plants. These salts are capable of storing heat at elevated temperatures (300–600°C) with low thermal losses, which makes it possible to store energy for long durations and generate power even after sunset [6]. Rock and concrete storage also uses solid materials like rocks, concrete, and ceramic bricks, which have high thermal stability. These materials are highly efficient in high-temperature applications like industrial waste heat recovery and solar furnaces. SHS is inexpensive and easy to store; however, the storage capacity is restricted by the specific heat and temperature range of the material.

1.6.2 Latent Heat Storage (LHS)

Latent heat storage systems use phase change materials (PCMs) to store and release energy in a phase transition, for example, solid-to-liquid or liquid-to-gas. The main benefit of PCMs is their high energy storage density and isothermal operation and is thus well suited for constant temperature storage applications. Organic PCMs, such as paraffin waxes and fatty acids, are most frequently applied for low-temperature uses (below 100°C) like building heating and cooling, solar dryers, and refrigeration. Inorganic PCMs, like salt hydrates and metallic PCMs, possess greater thermal conductivity and stability and are well suited for industrial thermal energy storage (TES) applications, such as solar power plants and waste heat recovery. To upgrade the efficiency of heat transfer and protect against leakage of material, encapsulated PCMs tend to be housed in metal or polymer capsules, enhancing their performance and longevity overall. LHS offers greater energy storage density compared to SHS but needs efficient encapsulation and heat transfer processes for optimized energy absorption and release [15].

1.6.3 Thermochemical Energy Storage (TCES)

Thermochemical energy storage (TCES) is a next-generation thermal energy storage (TES) technology that stores energy in reversible chemical reactions. In contrast to sensible heat storage (SHS) and latent heat storage (LHS), which are based on temperature change or phase change, TCES stores energy in chemical bonds, providing much higher storage densities and long-duration storage with no thermal losses. Adsorption storage employs zeolites and silica gel, among other materials, to capture heat through adsorbing water vapor, making it ideal for building heating and dehumidification purposes. Chemical reaction storage uses materials that experience endothermic and exothermic reactions to store and release heat. Metal oxides ($CaO/Ca(OH)_2$) and salt hydrates, for example, are used extensively in high-temperature TES for industrial use. In addition, hydrogen-based TES facilitates thermal energy storage through solar reactor-based thermochemical hydrogen production, providing an attractive strategy for long-term renewable energy storage. TCES provides more energy density and longer storage period than SHS and LHS but needs enhanced material development and system integration to become applicable on a large scale [16].

1.7 ADVANCEMENTS IN SOLAR THERMAL ENERGY TECHNOLOGIES

Solar thermal energy technology has experienced drastic improvements in efficiency, sustainability, and usability across industries in recent years. While the world's energy demand keeps surging, having cleaner and efficient sources of energy has become pivotal. Solar thermal energy provides an encouraging solution that utilizes the energy of the sun's heat in a range of applications, such as electricity generation, industrial processes, and domestic heating. They address improvements in capturing solar energy, efficiency of heat transfer, and coupling with other sources of renewable energy for good performance [17]. Some of the developments here are as follows.

1.7.1 High-Efficiency Materials and Coatings

One of the main fields of development in solar thermal technology is the creation of high-efficiency materials and selective coatings. These materials are created to maximize the absorption of solar radiation while reducing thermal losses. Selective coatings, including black nickel, titanium-based films, and carbon nanotubes, increase the efficiency of solar collectors to trap heat and increase overall efficiency. Also, nanotechnology advances have made it possible to develop nanostructured coatings that enhance thermal conductivity, decrease emissivity, and enhance durability of solar absorbers at elevated temperatures [17].

Recent advances in ceramic-based coatings have also been responsible for increased efficiency in solar thermal systems. Such coatings are very resistant to high temperatures and are more efficient under extreme weather conditions, hence are applicable to concentrated solar power (CSP) plants. The other key innovation is the application of anti-reflective and hydrophobic coatings that repel dust and moisture on solar collectors. This minimizes maintenance and increases long-term efficiency [17].

In addition, developments in aerogel-based materials have demonstrated high potential in minimizing heat losses in solar collectors. Ultra-lightweight and highly insulating aerogels are being considered as thermal insulators in solar receivers to reduce heat dissipation and enhance overall system efficiency [17].

1.7.2 Innovations in Heat Transfer Fluids

Heat transfer fluids (HTFs) are important in solar thermal systems in that they aid in transferring absorbed heat from the solar collectors to storage or for direct use. Conventional HTFs, for example, water, oils, and molten salts, lack thermal stability and efficiency at high temperatures. Recent studies have been directed toward the synthesis of nanofluids—nanoparticle-loaded HTFs with aluminum oxide, copper oxide, and carbon nanotubes that have higher thermal conductivity and specific heat capacity than traditional fluids [17].

Nanofluids enhance heat transfer effectiveness through the reduction of viscosity and enhancement of thermal diffusivity, resulting in improved energy absorption and distribution within solar thermal systems. Phase change materials (PCMs) have also been combined with HTFs to facilitate thermal energy storage, providing a more stable and consistent supply of energy even during the absence of sunlight. PCMs store surplus heat during the peak sun hours and release it when energy is needed more, enhancing the efficiency of solar thermal plants [17].

The other significant achievement in HTFs is the invention of ionic liquids, which provide better thermal stability and non-volatility than traditional fluids. These new-generation fluids have proved to be promising in enhancing the operating temperature range of solar thermal power plants, especially in CSP systems [17].

1.7.3 Hybrid Solar Thermal Systems

Hybrid solar thermal systems combine solar thermal technology with other renewable sources, including photovoltaic (PV) panels, biomass, and geothermal power, to provide higher reliability and total energy output. Among the most exciting innovations in this regard are photovoltaic-thermal (PVT) systems, which blend PV panels with solar thermal collectors to produce both electricity and heat at the same time [18]. Hybrid solar thermal systems have already been put to effective use in homes and business establishments to supply hot water and power, achieving maximum energy efficiency. For example, solar-assisted heat pump systems use solar thermal power to pre-heat water, minimizing the electrical load on heat pumps and enhancing overall efficiency [19].

In industrial uses, hybrid CSP-biomass systems have been of interest as a renewable energy option. Hybrid systems employ biomass as a reserve energy source to maintain operation even during low solar radiation conditions. By combining CSP with biomass gasification or combustion technologies, hybrid systems can offer stable and dispatchable power generation while minimizing fossil fuel reliance [19].

Another significant innovation is the coupling of solar thermal energy with desalination technologies. Solar heat is employed in solar desalination systems to power thermal desalination technologies like multi-effect distillation (MED) and multi-stage flash (MSF) to achieve a long-term solution for freshwater supply in water-scarce areas [19].

1.8 COMPUTATIONAL INTELLIGENCE AND SMART TECHNOLOGIES IN SOLAR THERMAL SYSTEMS

The growth in computational intelligence and intelligent technologies has fundamentally revolutionized solar thermal systems, rendering them efficient, dependable, and adaptive to changing environmental conditions. These technologies apply artificial intelligence (AI), machine learning (ML), optimization algorithms, and the Internet of Things (IoT) to enhance the performance of solar thermal uses. By combining these intelligent solutions, solar thermal systems can forecast energy requirements, optimize thermal energy storage, automate service, and increase integration with smart grids and microgrids. This section discusses some of the computational intelligence approaches and smart technologies that are useful in the creation of next-generation solar thermal systems [20].

Figure 1.7 provides an overview of intelligent technologies in solar thermal systems, identifying four main areas: artificial intelligence (AI), machine learning (ML), optimization techniques, and the Internet of Things (IoT). AI improves predictive ability and system responsiveness to enable enhanced decision-making and performance. ML aims at optimizing energy management and storage solutions, enhancing overall efficiency. Optimization techniques also improve system reliability and effectiveness by optimizing operational processes. In the meantime, IoT provides easy integration with

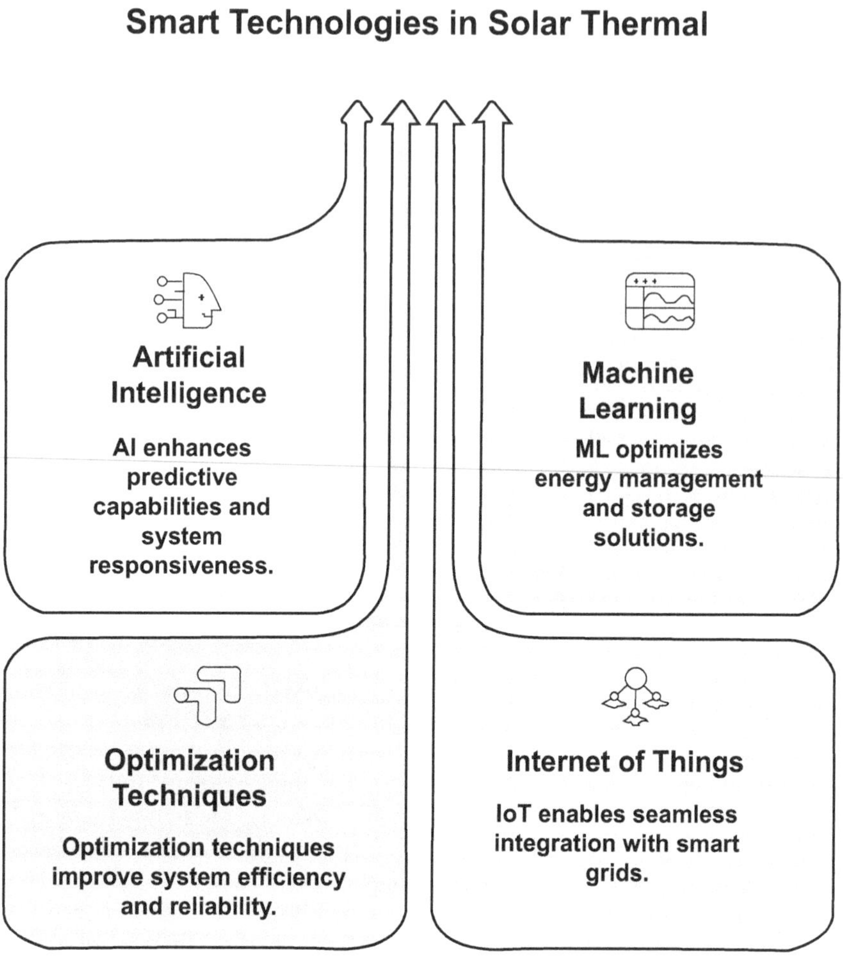

Figure 1.7 Smart technologies in solar thermal.

smart grids for improved connectivity and communication among the various components of the system. The diagram shows how the technologies as a whole help bring solar thermal energy solutions forward.

1.8.1 Neural Networks, Genetic Algorithms, and Fuzzy Logic

Neural networks, genetic algorithms, and fuzzy logic are important in improving the efficiency and adaptability of solar thermal systems. Neural networks (NNs) are AI-based models of the human brain employed for pattern recognition, predicting solar energy output, anticipating system failure, and optimizing heat transfer efficiency. With the analysis of large datasets, neural networks enhance energy output predictions to enable operators to dynamically adjust system parameters to achieve improved performance [21]. Genetic algorithms (GAs) replicate natural selection to identify the optimum settings for collector orientation, working fluid selection, and heat exchanger design. GAs iteratively evolve solutions to optimize energy efficiency while reducing operational costs. Fuzzy logic, as opposed to binary logic, is a more flexible means of addressing uncertainties in solar thermal systems. It is best applied in dealing with variable weather patterns where rule-based conventional controls can prove inadequate. Fuzzy logic controllers provide smooth and adaptive operation by modulating system parameters like temperature, flow rate, and storage conditions through approximate reasoning [22]. By integrating these computational intelligence methods, solar thermal systems can realize enhanced adaptability, robustness, and operation efficiency even with changing environmental conditions [23, 24].

1.8.2 Machine Learning for Predictive Modeling

Machine learning (ML) methods are extensively applied in solar thermal systems for predictive modeling, which improves performance forecasting, fault detection, and decision-making. Predictive energy modeling employs ML algorithms to examine past solar radiation data, weather conditions, and system performance indicators to predict energy generation and consumption [25–28]. These models optimize energy storage and distribution to address changing demands effectively. Machine learning methods, including decision trees, support vector machines (SVMs), and deep learning, are employed to identify anomalies in solar thermal systems. Through the detection of deviations from normal behavior, ML facilitates early fault detection and avoids system failure. Sophisticated ML methods like reinforcement learning enable solar thermal systems to self-optimize by continuous learning. Through interaction with the environment, such systems can modify operation parameters to optimize heat uptake and minimize losses of energy. Application of machine learning to solar thermal processes enhances efficiency, decreases downtime, and overall improves reliability [29–33].

1.8.3 Optimization Algorithms and Strategies for Efficiency

Optimization algorithms are the most important elements in increasing the efficiency of solar thermal systems through the reduction of energy losses and enhancement of thermal performance. Metaheuristic optimization methods like Particle Swarm Optimization (PSO), Ant Colony Optimization (ACO), and Simulated Annealing (SA) are applied for finding the optimal system parameters, such as heat exchanger sizes, fluid flow rates, and coatings on the absorber. They are capable of finding nearly optimal solutions at high speeds, even for highly complex, nonlinear problems. Mathematical and computational models, such as linear and nonlinear programming methods, aid in the optimization of solar thermal plant operation. The models aid in resource allocation, scheduling energy storage, and optimizing heat transfer efficiency [34–36]. Hybrid optimization methods that integrate several methods, like genetic algorithms and neural networks, enhance system performance. For instance, these methods can optimize solar tracking systems to ensure maximum solar absorption during the day. Through implementation of these optimization techniques, solar thermal systems are capable of having better energy conversion efficiency, lower cost of operation, and enhanced long-term viability [37–39].

1.8.4 IoT and Sensor Networks for Monitoring and Control

The confluence of the Internet of Things (IoT) and sensor networks has transformed real-time monitoring and control in solar thermal systems. IoT-enabled smart sensors gather temperature, pressure, intensity of radiation, and heat flow data in real-time, offering instant insights into system performance. Sensors wirelessly communicate with cloud-based platforms, facilitating remote monitoring and automation control. The large volumes of data produced by IoT devices are analyzed through big data analytics and cloud computing to enable predictive maintenance, performance optimization, and fault detection in advance. Remote control and automation technologies leverage IoT to facilitate real-time adjustments of solar thermal plant operations [40, 41]. Dynamic control of flow rates, heat storage levels, and collector angles can be managed by automated controllers to maximize energy absorption and thermal efficiency. The fusion of sensor networks and IoT converts solar thermal systems into efficient, adaptive, and intelligent energy technologies, minimizing the scope for manual interference while maintaining optimum performance [42].

Figure 1.8 shows the IoT-based process of optimizing solar thermal systems, marking an ongoing cycle of principal functions. It starts with data acquisition through intelligent sensors, then data transfer to cloud platforms. The information gathered is processed for analysis and predictions, allowing remote observation of system performance. Automated control finally takes care of real-time adjustments, resulting in performance optimization by increasing energy absorption and efficiency [43].

1.8.5 Predictive Maintenance and Fault Detection

Predictive maintenance and fault detection methodologies are critical for minimizing downtime and maximizing the solar thermal system's life.

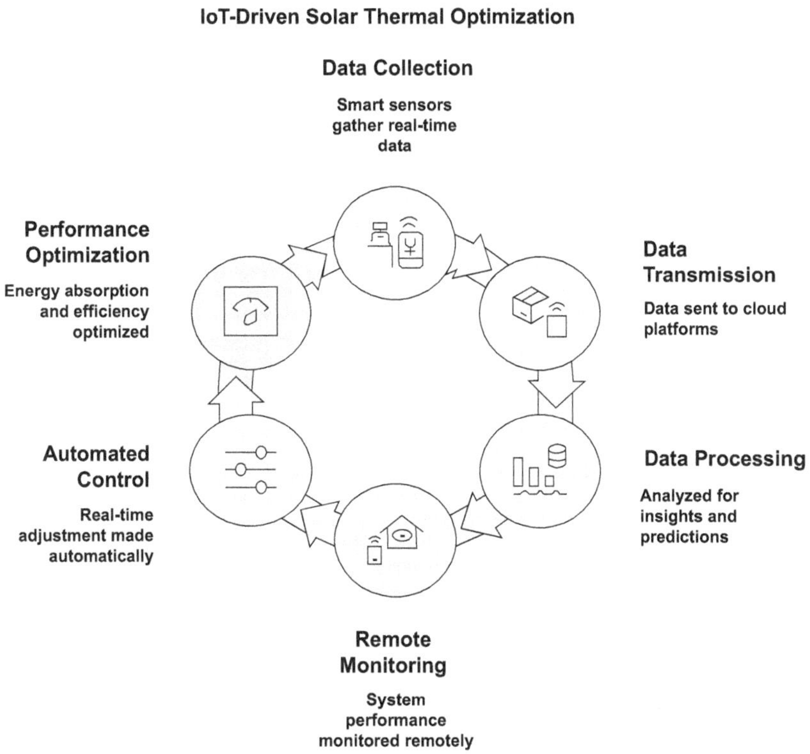

Figure 1.8 IoT-based solar thermal optimization.

AI-maintenance strategies based on machine learning algorithms process historic operational data and predict when a component will likely fail. By doing so, it enables planned maintenance prior to failures, and this minimizes repair expenses as well as system downtime. Sophisticated sensing methods, such as vibration analysis, thermal imaging, and acoustic monitoring, assist in the detection of problems like leaks, overheating, and structural flaws in collectors and heat exchangers. Automated fault detection systems continuously monitor system performance, detecting inefficiencies and sending real-time notifications to operators. These automated diagnostic tools reduce energy losses and provide constant system output. Predictive maintenance practices improve reliability, maximize system availability, and lower the operational expenses of solar thermal systems [44].

1.8.6 Integration with Smart Grids and Microgrids

Solar thermal systems integration with smart grids and microgrids improves energy distribution, stability, and resilience. AI-based smart grids enable solar thermal systems to harmonize with the energy grid, adjusting supply and demand seamlessly. This integration minimizes power fluctuations and ensures energy distribution is always guaranteed. Microgrids provide localized solar power generation and storage, which lowers reliance on central power grids and enhances energy security, particularly in remote or off-grid locations. Smart grids also provide bidirectional energy exchange, enabling surplus thermal energy to be returned to the grid during over-supply compared to demand. Stored energy can be withdrawn from the grid when there is low solar radiation to ensure stable operation. Through their integration with smart grids and microgrids, solar thermal systems make a resilient and sustainable energy infrastructure possible, which guarantees efficient management of energy both locally and globally [45, 46].

1.9 APPLICATIONS AND FUTURE PROSPECTS

Solar thermal energy technologies are being utilized across different industries, such as power generation, industrial process heat, and space heating. The use of concentrated solar power (CSP) plants has grown, with nations such as Spain, the United States, and China investing in large-scale solar thermal power plants. These initiatives are intended to utilize the vast solar resource to produce electricity in a sustainable and environmentally friendly way. CSP technologies such as parabolic troughs, solar towers, linear Fresnel reflectors, and dish Stirling systems are being designed for enhanced efficiency and cost-effectiveness. Emerging technologies like molten salt storage and supercritical CO_2 cycles are being investigated for improved efficiency and storage in CSP systems. Molten salt storage allows solar thermal power plants to store surplus heat and produce electricity even during periods of no sunlight, thus enhancing grid stability and providing a secure power supply. Likewise, supercritical CO_2 cycles are increasingly being considered as a viable alternative to conventional steam-based power cycles with greater thermal efficiency and lower water usage.

In industrial applications, solar thermal energy is being utilized more and more for process heat in food processing, textile, and chemical industries. The capability to supply high-temperature heat renders solar thermal energy a good alternative to traditional fossil-fuel-fired heat systems. Industries with a need for continuous high-temperature heat, for example, metal processing and cement production, are also looking into the potential for incorporating solar thermal energy to minimize their carbon footprint and operating expenses. Further, hybrid systems that include solar thermal energy and other renewable energy sources like biomass or geothermal are being developed to provide an uninterrupted supply of energy in industrial processes.

Solar thermal technologies are also gaining popularity in the residential sector for space heating and water heating. Solar water heaters, with either flat-plate collectors or evacuated tube collectors, are becoming increasingly popular among residential and commercial consumers as a cost-effective and energy-efficient alternative to electric or gas-based systems. Solar thermal power-driven district heating systems are also in the making in colder climates, providing a clean alternative to heating big populations with lower fossil fuel reliance. Future growth and advancement in sophisticated materials, energy storage, and hybrid integration will continue to play a vital part in advancing solar thermal technology further. The use of smart monitoring and control systems, including artificial intelligence (AI) and Internet of Things (IoT) technologies, will be used to improve system performance and maximize energy output. AI-based predictive maintenance will enable the system to detect potential failures beforehand, minimizing downtime and maintenance expenses. IoT sensors can give real-time information about system performance, making remote monitoring and automatic adjustments possible to maximize efficiency.

Furthermore, government and international organization policy support and financial incentives are anticipated to spur additional solar thermal energy adoption. Subsidies, tax credits, and supportive regulatory policies can incentivize businesses and residents to invest in solar thermal technology, speeding the shift toward a cleaner energy future. As the global fight against climate change increases, solar thermal energy is likely to be an important contributor to curbing greenhouse gas emissions and ensuring energy security globally.

REFERENCES

[1] P. Ray, "Renewable energy and sustainability," *Clean Technol. Environ. Policy*, vol. 21, pp. 1517–1533, 2019.

[2] M. Imran Khan, F. Asfand, and S. G. Al-Ghamdi, "Progress in research and technological advancements of commercial concentrated solar thermal power plants," *Solar Energy*, vol. 249, pp. 183–226, January 2023, https://doi.org/10.1016/j.solener.2022.10.041.

[3] T. Kurbatova and T. Perederii, "Global trends in renewable energy development," In *2020 IEEE KhPI Week on Advanced Technology (KhPIWeek)*, pp. 260–263, October 2020, https://doi.org/10.1109/khpiweek51551.2020.9250098.

[4] M. A. Rosen, "Renewable energy and energy sustainability," In *Design and Performance Optimization of Renewable Energy Systems*, pp. 17–31, 2021, https://doi.org/10.1016/b978-0-12-821602-6.00002-x.

[5] D. Y. Goswami, *Principles of Solar Engineering*, CRC Press, 2015, https://doi.org/10.1201/b18119.

[6] Z. Said, A. A. Hachicha, S. Aberoumand, B. A. A. Yousef, E. T. Sayed, and E. Bellos, "Recent advances on nanofluids for low to medium temperature solar collectors: energy, exergy, economic analysis and environmental impact," *Prog.*

Energy Combust. Sci., vol. 84, p. 100898, May 2021, https://doi.org/10.1016/j.pecs.2020.100898.

[7] F. B. Mitri and K. R. Anderson, "Comparison of solar thermal collectors and solar PV heat pump systems," In *ASME 2022 16th International Conference on Energy Sustainability*, July 2022, https://doi.org/10.1115/es2022–81560.

[8] J. A. Duffie and W. A. Beckman, *Solar Engineering of Thermal Processes*, John Wiley & Sons, 2013, https://doi.org/10.1002/9781118671603.

[9] T. L. Bergman, *Fundamentals of Heat and Mass Transfer*, John Wiley & Sons, Hoboken, NJ, 2011.

[10] G. C. Vliet, "Principles of solar engineering," *J. Solar Energy Eng.*, vol. 122, no. 2, pp. 114, May 2000, https://doi.org/10.1115/1.1288930.

[11] S. A. Kalogirou, *Solar Energy Engineering: Processes and Systems*, Elsevier, 2023.

[12] "Solar energy," *Sustain. Energy*, pp. 251–298, January 2023, https://doi.org/10.1017/9781009043113.012.

[13] H. L. Zhang, J. Baeyens, J. Degrève, and G. Cacères, "Concentrated solar power plants: Review and design methodology," *Renew. Sustain. Energy Rev.*, vol. 22, pp. 466–481, June 2013, https://doi.org/10.1016/j.rser.2013.01.032.

[14] V. V. Tyagi, S. C. Kaushik, and S. K. Tyagi, "Advancement in solar photovoltaic/thermal (PV/T) hybrid collector technology," *Renew. Sustain. Energy Rev.*, vol. 16, no. 3, pp. 1383–1398, April 2012, https://doi.org/10.1016/j.rser.2011.12.013.

[15] I. Sarbu and C. Sebarchievici, "A comprehensive review of thermal energy storage," *Sustainability*, vol. 10, no. 1, p. 191, January 2018, https://doi.org/10.3390/su10010191.

[16] "Sensible thermal energy storage systems," In *Thermal Energy Storage*, pp. 37–80, March 2023, https://doi.org/10.1142/9789811271182_0002.

[17] Toxigon, *Advances in Solar Technology: 2024 Guide*, 2024, https://toxigon.com/advances-in-solar-technology-2024.

[18] Ashwani Kumar, Nitesh Dutt, and Mukesh Kumar Awasthi, *Heat Transfer Enhancement Techniques: Thermal Performance, Optimization and Applications*, John Wiley & Sons, Inc., USA, 2024, https://doi.org/10.1002/9781394270996.

[19] Sugimat, *Sugimat's Thermal Installation for the First Hybrid Solar-Thermal-Biomass Plant at Full Capacity*, 2023, https://www.sugimat.com/en/sugimats-thermal-installation-for-the-first-hybrid-solar-thermal-biomass-plant-at-full-capacity.

[20] S. Nesmachnow and C. Risso, "A computational intelligence approach for solar photovoltaic power generation forecasting," *Renew. Energies*, vol. 2, no. 1, January 2024, https://doi.org/10.1177/27533735241237990.

[21] S. A. Kalogirou, E. Mathioulakis, and V. Belessiotis, "Artificial neural networks for the performance prediction of large solar systems," *Renew. Energy*, vol. 63, pp. 90–97, March 2014, https://doi.org/10.1016/j.renene.2013.08.049.

[22] B. Maroua, Z. Laid, H. Benbouhenni, Z. M. S. Elbarbary, I. Colak, and M. M. Alammer, "Genetic algorithm type 2 fuzzy logic controller of microgrid system with a fractional-order technique," *Scient. Rep.*, vol. 15, no. 1, February 2025, https://doi.org/10.1038/s41598-025-90239-1.

[23] P. Tahmasebi and A. Hezarkhani, "A hybrid neural networks-fuzzy logic-genetic algorithm for grade estimation," *Comput. Geosci.*, vol. 42, pp. 18–27, May 2012, https://doi.org/10.1016/j.cageo.2012.02.004.
[24] A. Muhammad Ibrahim, S. M. Lawan, R. Abdulkadir, N. S. Shuaibu, M. Uzair, M. G. Indabawa, M. Ibrahim, and A. M. Aliyu, "Solar radiation prediction using an improved adaptive neuro-fuzzy inference system (ANFIS) optimization ensemble," In *Adaptive Neuro-Fuzzy Inference System as a Universal Estimator*, March 2024, https://doi.org/10.5772/intechopen.1003891.
[25] Prashis Raghuwanshi, "Predictive modeling and forecasting of solar power generation using machine learning techniques," *J. Adv. Scholar. Res. Allied Educ.*, vol. 21, no. 5, pp. 207–213, July 2024, https://doi.org/10.29070/zccv5324.
[26] M. Nayal, A. K. Sharma, S. Jain, and V. P. Singh, "Optimization techniques of solar thermal and hybrid energy systems," In *Highly Efficient Thermal Renewable Energy Systems*, pp. 104–120, April 2024, https://doi.org/10.1201/9781003472629–7.
[27] P. Khobragade, P. Ghutke, V. P. Kalbande, and N. Purohit, "Advancement in internet of things (IoT) based solar collector for thermal energy storage system devices: A review," In *2022 2nd International Conference on Power Electronics & IoT Applications in Renewable Energy and its Control (PARC)*, pp. 1–5, January 2022, https://doi.org/10.1109/parc52418.2022.9726651.
[28] S. Saravanan, P. Pandiyan, T. Rajasekaran, N. Prabaharan, T. Chinnadurai, and R. Tiwari, "Machine learning-based predictive maintenance for solar plants for early fault detection and diagnostics," *Photovolt. Syst.*, pp. 85–103, February 2022, https://doi.org/10.1201/9781003202288–5.
[29] Y. Methkal Abd Algani, V. Sreenivasa Rao, and R. Saravanakumar, "AI-powered secure decentralized energy transactions in smart grids: Enhancing security and efficiency," In *2024 IEEE 3rd International Conference on Electrical Power and Energy Systems (ICEPES)*, pp. 1–5, June 2024, https://doi.org/10.1109/icepes60647.2024.10653598.
[30] Arijit Kundu, Ashwani Kumar, Mukesh Kumar Awasthi, Gaurav Dwivedi, and Anoop Kumar Shukla, "An explorative study of earth-air heat exchanger for sustainable building applications," *Int. J. Energy Clean Environ.*, vol. 26, no. 3, pp. 1–22, 2025, https://doi.org/10.1615/InterJEnerCleanEnv.2024052513.
[31] Zahrapanah Razaviyn, Milad Heidari, Sivasakthivel Thangavel, Vikas Verma, Ashwani Kumar, and Ashok Kumar Yadav, "Marine energy harvesting from fluid flow via vortex induced vibrations," *Int. J. Thermofluids*, vol. 25, p. 101015, 2025, https://doi.org/10.1016/j.ijft.2024.101015.
[32] Dan Singh Pimoli, Pritosh Tomar, Ashwani Kumar, Gaurav Dwivedi, and Anoop Kumar Shukla, "Experimental investigation of friction factor and heat transfer enhancement using circular perforated twisted tape inserts in heat exchangers," *Int. J. Energy Clean Environ.*, 2025, https://doi.org/10.1615/InterJEnerCleanEnv.2024053616.
[33] M. Mokhtari, S. Jamei, Z. Razaviyan, M. Heidari, S. Thangavel, and Ashwani Kumar, "Numerical and hydrodynamic analysis of a container vessel due to the change of collision wave angle at different speeds in regular wave," *J. Mech. Eng. Sci.*, vol. 18, no. 4, pp. 10247–10260, 2024, https://doi.org/10.15282/jmes.18.4.2024.3.0809.

[34] Pritosh Tomar, Prateek Chauhan, Ashwani Kumar, and Anoop Kumar Shukla, "Numerical investigation of thermal performance enhancement of solar reservoir using flash cycle," *J. Adv. Res. Fluid Mech. Therm. Sci.*, vol. 123, no. 1, pp. 197–221, 2024, https://doi.org/10.37934/arfmts.123.1.197221.

[35] Nitesh Dutt, Ankush Hedau, Ashwani Kumar, Mukesh Kumar Awasthi, and Varun Pratap Singh, "Effect of duct length variation on solar air heater performance for smooth and D-shaped roughened absorber plate," *Heat Transf.*, 2024, https://doi.org/10.1002/htj.23115.

[36] Yadav Dhananjay, Mukesh Kumar Awasthi, Ashwani Kumar, and Nitesh Dutt, "Swirling capillary instability of Rivlin–Ericksen liquid with heat transfer and axial electric field," *Physics*, vol. 6, no. 2, pp. 828–844, 2024, https://doi.org/10.3390/physics6020051.

[37] Nitesh Dutt, Ankush Hedau, Ashwani Kumar, Mukesh Kumar Awasthi, Sachin Hedau, and Chandan Swaroop Meena, "Thermo-hydraulic performance investigation of solar air heater duct having staggered D-shaped ribs: Numerical approach," *Heat Transf.*, vol. 53, no. 3, pp. 1501–1531, 2024, https://onlinelibrary.wiley.com/doi/10.1002/htj.22998.

[38] Ashwani Kumar, Pratibha S. Agrawal, and Yogesh Kumar Singla, *Sustainable Development With Carbon Capture, Utilization and Storage Technology*, Springer Nature Switzerland, 2025.

[39] Dhananjay Yadav, Mukesh Kumar Awasthi, and Ashwani Kumar, *Next Generation Renewable Thermal Energy Harvesting, Conversion & Storage Technologies*, Elsevier, 2025.

[40] Ashwani Kumar, Arbind Prasad, and Gaurav Kumar, *Sustainable Mobility: Policies, Challenges and Advancements*, John Wiley & Sons, Inc., 2024, https://doi.org/10.1002/9781394166831.

[41] Tikendranath Verma, Thokchom Subhaschandra Singh, Upendra Rajak, Prerana Nashine, Gaurav Dwivedi, and Ashwani Kumar, *Clean Energy: Technology, Advances and Applications*, CRC Press, Boca Raton, FL, 2024, https://doi.org/10.1201/9781003521341.

[42] Payal Bansal, Rajeev Kumar, Ashwani Kumar, and Daniel Dasig Jr., *Artificial Intelligence and Communication Techniques in Industry 5.0*, CRC Press, Boca Raton, FL, 2024, https://doi.org/10.1201/9781003494027.

[43] Abhishek Kumar, Avinash Kumar, and Ashwani Kumar, *Performance Characterization of Lubricants*, CRC Press, Boca Raton, FL, 2024, https://doi.org/10.1201/9781003535447.

[44] Chandan Swaroop Meena, Ashwani Kumar, Varun Pratap Singh, and Aritra Ghosh, *Sustainable Technologies for Energy Efficient Buildings*, CRC Press, Boca Raton, FL, 2024, https://doi.org/10.1201/9781003496656.

[45] Ashwani Kumar, Varun Pratap Singh, Chandan Swaroop Meena, and Nitesh Dutt, *Thermal Energy Systems: Design, Computational Techniques and Applications*, CRC Press, Boca Raton, FL, 2023, https://doi.org/10.1201/9781003395768.

Chapter 2

Application of Machine Learning in Greenhouse-Type Solar Dryer Systems

A Review

Shubham Mishra and Sachin Kumar Nikam

2.1 INTRODUCTION

Greenhouse-type solar dryers are widely used for drying agricultural products, herbs, and other materials [1, 2]. Traditional drying methods face challenges such as dependency on weather conditions and suboptimal efficiency [3, 4]. Researchers in the past worked on the greenhouse solar dryer systems using various experimental design approaches [1–10].

In a greenhouse dryer, working under forced convection mode was found to be best for high moisture content crops; however, the greenhouse dryer under natural convection mode was found to be best for low moisture content crops [1]. Crops dried in the greenhouse dryer are of a superior quality and color as compared to open sun drying [1, 2]. Color, quality, taste, and nutritional value of the dried product are better in greenhouse solar drying than open sun drying [2]. Sahu et al. [8] studied solar drying methods in greenhouses, focusing on the use of dryers to control moisture. Forced convection drying is optimal for high moisture content, while natural convection dryers are suitable for low moisture content [8]. Photovoltaic-integrated solar dryers are ideal for commercial use in isolated locations [8]. Kaewkiew et al. [10] investigated the performance of a large-scale greenhouse solar dryer by drying chili in Thailand. They found that reduced drying time and better color of dried chilis were achieved by drying them using greenhouse solar dryer as compared to open natural solar drying [10].

A large-scale greenhouse solar dryer with an LPG burner has been developed for drying osmotically dehydrated tomatoes [3]. This method significantly reduces drying time compared to open-air sun drying, resulting in high-quality dried products [3]. The dryer also addresses the issue of drying interruption by rain and cloudy periods [3]. Tiwari et al. [4] reviewed greenhouse solar dryer and found that thin drying is better as compared to deep bed drying due to better and uniform moisture removal rate [4]. In the context of the development of greenhouse solar dryer for medicinal plants and herbs, Muller et al. [5] found that greenhouse significantly increases

DOI: 10.1201/9781003634737-2

profitability by reducing drying time and allowing plant production to continue even after the removal of absorber tissues [5].

According to Koyuncu T.'s [6] investigation on the effectiveness of green solar dryer systems, greenhouse solar dryers—which are generally thought to be of higher quality and efficiency—are two to five times more effective than open sun dryers. In another investigation, Saini et al. [7] highlighted the importance of CIGS PV technology in integrating solar energy into greenhouse systems, highlighting its energy payback time and reduced greenhouse gas emissions. The c-Si option remains the optimal choice for optimal energy use in rural areas [8]. Patil and Gawande's [9] study on solar tunnel greenhouse systems revealed that PV-ventilated solar tunnel greenhouse dryers offer a reliable and intelligent alternative to traditional sun drying methods, but their operation requires optimization. They emphasize the need for efficient and effective solar energy solutions [10].

In addition to experimental investigations, researchers also worked in the recent past on prediction of greenhouse solar dryer systems using machine learning approaches [11–15]. Gaussian process regression (GPR), artificial neural network (ANN), decision tree, K-nearest neighbor, support vector machine, gradient boosted regression trees, random forest, multiple linear regression, histogram gradient boosting, Bayesian ridge regression, extreme gradient boosting (XGBoost), light gradient boosting (LightGBM), unscented Kalman filter were found to be major machine learning algorithms that were employed in greenhouse solar dryer systems [16–20]. The objective of the present research work is to review the application of machine learning techniques in design, modeling, and analysis of greenhouse solar dryer systems [21–25].

2.2 GREENHOUSE-TYPE SOLAR DRYER SYSTEMS: AN OVERVIEW

Greenhouse-type solar dryers trap heat inside a closed structure to use solar energy for drying. They are made up of essential parts, including fans to control humidity and temperature, collectors, and clear coverings [26–28]. Although they lower operating costs and energy reliance, climate fluctuations have a significant impact on their efficiency. Greenhouse basically operates in either passive (natural convection) or active mode (forced convection). The classification of greenhouse solar dryers according to our literature analysis is displayed in Figure 2.1 [1, 2].

Greenhouse dryers can be passive or active, with passive mode allowing air to circulate naturally [29, 30]. Active dryers have an exhaust fan for moist air transfer. Solar dryers include PV-integrated greenhouses, solar air heaters, inclined north walls, and thermal storage materials like sand, rock-bed, black concrete floor, and PVC sheet. These dryers have been developed to

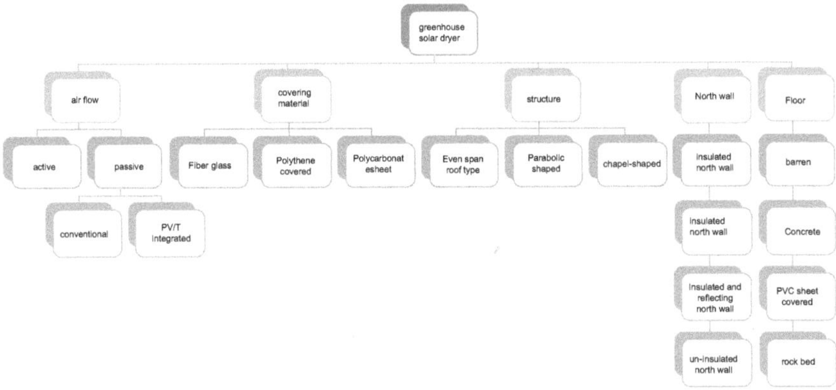

Figure 2.1 Classification of greenhouse solar dryer based on our literature review [2].

capture maximum radiation. Roof-even dryers and dome-shaped dryers are two types of greenhouse dryers, each with unique structural characteristics. Dome-shaped dryers aim to maximize sun radiation use, while roof-even dryers ensure proper air mixing. These dryers are divided into passive and active modes based on heat transfer mechanisms, ensuring efficient use of sunlight [31–34].

The thermosyphic effect is a key factor in the operation of greenhouse dryers, which utilize both passive and forced convection modes [35–37]. The passive mode involves the use of a chimney or roof ventilator, while the forced convection mode involves the use of an exhaust fan to ventilate humid air [1–4, 8, 9, 25]. Sun drying, the oldest method for drying agricultural products, has drawbacks like spoiling from wind, dust, rain, storms, and insects, and requires significant space and time to complete [9]. Open sun drying, a method of drying crops, can lead to product decolorization due to pollution, prompting the development of closed drying methods [4, 6]. Table 2.1 gives the advantages of greenhouse solar dryer over open sun drying. It may be concluded on the basis of previous discussion that greenhouse solar dryer was found to be more efficient than open drying systems in terms of reduced drying time, reduced chances cracking, and superior thermal performance [38–40].

2.3 MACHINE LEARNING IN SOLAR DRYING SYSTEMS

Machine learning is widely used in engineering tasks like regression and classification, learning from a dataset and making predictions using its acquired knowledge [13, 14]. Predictive modeling technology is utilized to evaluate the compatibility of dependent and independent factors [15]. The process of modeling, simulating, and predicting solar drying is a complex and time-consuming

Table 2.1 Advantages of Greenhouse Solar Dryer over Open Sun Drying

S. No	*Advantages of greenhouse solar dryer over open drying*	*Ref.*
1	Improvement in color, texture, and flavor; eliminates insect, microbe, and mycotoxin contamination, reduces drying time by 50%, minimizes drying and storage losses, and extends product life significantly.	[4, 8, 22]
2	Crop is shielded from direct sunlight, which prevents potential crop cracking issues.	[4, 8]
3	Compared to open-air drying, greenhouse dryers are 2.5 times more efficient. Additionally, using the chimney and black surface increases the dryers' efficiency.	[6]
4	A greenhouse drier reduces drying time by regulating temperature and relative humidity.	[6]
5	The indirect solar dryer offers superior production quality, commercial production capacity, and excellent thermodynamic performance among other solar dryer types.	[14, 22]

task due to its non-behavioral nature [41–43]. Contemporary machine learning techniques present a unique opportunity to provide a solution to this issue [19]. Figure 2.2 shows the various advantages of machine learning algorithms in the prediction of solar drying characteristics and performance.

Table 2.2 shows the different types of machine learning algorithms employed in greenhouse solar drying systems [13–24].

It was found that forecasting for different solar drying characteristics was found to be for drying of items such as mint, red chili, garlic clove, mushroom, banana, orange, and beetroot [44, 45]. Besides this, it was found that during regression analysis conducted by researchers, the major evaluation indices employed were coefficient of determination (R^2), the mean absolute error (MAE), and the root mean square error (RMSE) [13, 14, 16, 19, 21–23]. It was also recognized that in majority of machine learning forecasting analysis of solar drying systems characteristics, gradient boosting regression tree (GBRT) technique was found to be better in comparison to other ML techniques [11–24]. In summary, application of machine learning techniques/models/algorithms in solar drying systems helps in making systems more efficient and adaptive to environmental changes, reduces operational costs, and serves real-time monitoring and control.

2.4 CONCLUSIONS AND FUTURE SCOPE

The present investigation deals with the review of the greenhouse solar drying systems and application of machine learning models and algorithms in solar dryer systems. It was found that greenhouse solar dryer systems can

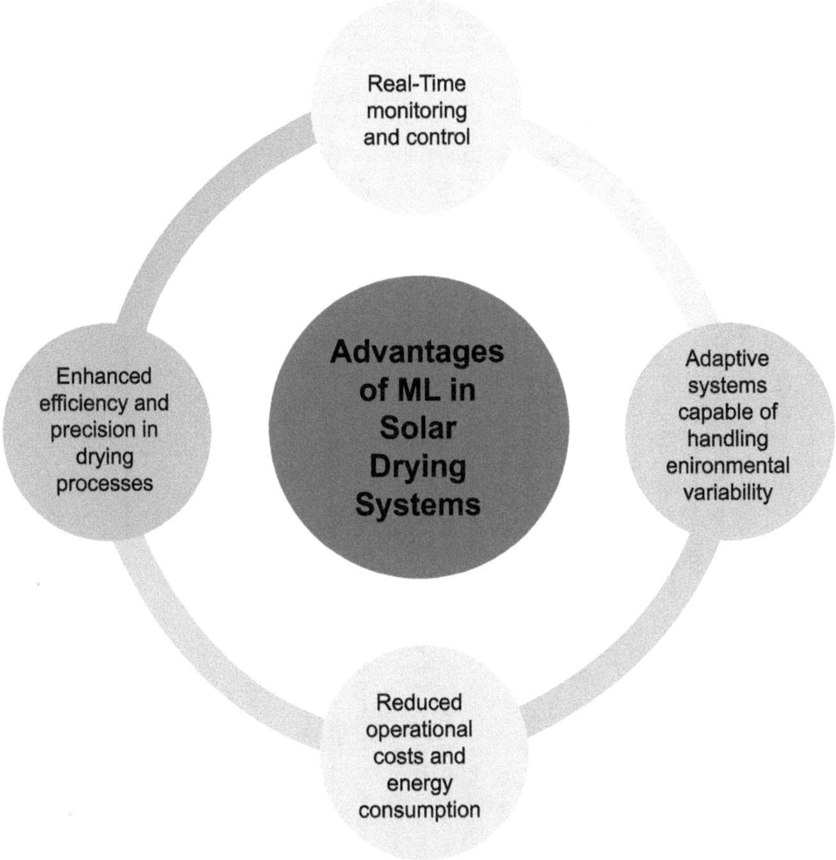

Figure 2.2 Advantages of ML in solar drying systems.

be classified on the basis of five major categories, namely airflow, covering material, structure, north wall, and floor types. Greenhouse solar dryer was found to be more efficient than open drying systems in terms of reduced drying time, reduced chances of cracking, and superior thermal performance. Machine learning techniques are utilized in solar drying systems to predict the features of these systems due to their nonlinear nature. A variety of food items such as mint, red chili, garlic clove, mushroom, banana, orange, and beetroot were utilized within solar dryer systems for prediction of characteristics of solar dryer systems. Coefficient of determination (R^2), the mean absolute error (MAE), and the root mean square error (RMSE) were found to be major evaluation indices during regression analysis. The gradient boosting regression tree (GBRT) technique was found to outperform other machine learning techniques in most machine learning forecasting analyses of solar drying system

Table 2.2 Applications of Machine Learning Techniques in Greenhouse Solar Dryer Systems.

Machine learning technique	*Purpose*	*Major inference*	*Ref.*
Artificial neural network (ANN), multilayer perceptron (MLP), radial bias function (RBF), Gaussian process regression (GPR)	Mass and temperature modeling of dried mint	Compared to other models, the radial bias function (RBF) model offers the best accuracy and the lowest MAPE.	[11]
Multilayer perceptron (MLP), decision tree (DT)	Modeling of heat transfer coefficient	The DT algorithm in GDNC was around 10% more accurate than MLP, and the MLP algorithm performed around 10% better than the DT in the GNFC.	[12]
Artificial neural network (ANN), k-nearest neighbor, support vector regression, gradient boosting regression tree, random forest	Performance forecasting of the drying operation of red chili	The gradient boosting regression tree model demonstrated superior performance in predicting the moisture content of red chili.	[13]
Decision trees (DT), gradient boosting regression (GBR), multiple linear regression (MLR), and random forest (RF)	Energy, exergy, environmental, and economic aspects of drying sweet potato	Random forest (RF) outperformed other ML algorithms during indirect solar dryer for drying sweet potato.	[14]
Gaussian processes (GP), support vector regression (SVR), elastic net, Bayesian ridge regression, linear regression, gradient boosting regression tree (GBRT)	Thermal performance of garlic cloves after drying in a heat control unit	Gradient boosting regression tree (GBRT) demonstrated good fit prediction performance.	[15]
Multiple linear regression, support vector machine, decision trees, bagging regressor, gradient boosting, Gaussian processes, Bayesian ridge regression, K-nearest neighbors, random forest, and artificial neural network	Solar dryer type, solar irradiance, ambient temperature, wind velocity, and relative humidity in tomato drying	The random forest method was the most successful in forecasting the solar dryers' outlet temperatures.	[16]

(Continued)

Table 2.2 (Continued)

Machine learning technique	*Purpose*	*Major inference*	*Ref.*
Artificial neural network (ANN)	Heat transfer coefficient during mushroom drying	The ANN model accurately predicts heat transfer coefficients, reducing mathematical computation effort and facilitating accurate determination of convective and evaporative heat transfer.	[18]
eXtreme gradient boosting (XGBoost) and light gradient boosting (LightGBM) algorithms	Drying characteristics of banana slices with an indirect solar dryer	XGBoost consistently performed better than LightGBM in prediction of drying characteristics.	[19]
Unscented Kalman filter	To dry orange slices using a direct-type solar dryer through natural convection	The unscented Kalman filter was used to experimentally learn the behavior of orange slices during drying, with a success rate of 0.93, and also during crop growth, demonstrating its effectiveness in learning.	[20]
Support vector regression, Bayesian ridge regression, elastic net, linear regression, and gradient boosting regression tree	Thermal performance of dual working medium solar energy drying system	Gradient boosting regression can be regarded as the best prediction performance	[21]
LightGBM model, Catboost model, recurrent neural network (RNN), and long short-term memory networks	An assessment of the material's characteristics, microstructure to enhance the beetroot via drying method.	Out of all the models examined, the Catboost model is the most accurate.	[22]
Singular spectrum analysis (SSA), convolutional neural networks (CNNs), and long short-term memory (LSTM) networks hybrid SSA-CNN-LSTM	Precise solar power forecasting	The study demonstrates the superior performance of SSA-CNN-LSTM, particularly in the 1-hour forward prediction window.	[23]

characteristics. Machine learning models and algorithms enhance solar drying systems' efficiency, adaptability to environmental changes, and real-time monitoring, thereby lowering operating costs. Future research in machine learning applications in solar drying systems includes developing hybrid models combining deep learning and traditional techniques, integrating IoT systems for autonomous operation, and conducting life cycle assessments of ML-enabled solar dryers to evaluate environmental impact.

REFERENCES

[1] Prakash, O. and Kumar, A., 2014. Solar greenhouse drying: A review. *Renewable and Sustainable Energy Reviews*, *29*, pp. 905–910.
[2] Singh, P., Shrivastava, V. and Kumar, A., 2018. Recent developments in greenhouse solar drying: A review. *Renewable and Sustainable Energy Reviews*, *82*, pp. 3250–3262.
[3] Janjai, S., 2012. A greenhouse type solar dryer for small-scale dried food industries: Development and dissemination. *International Journal of Energy and Environment*, *3*(3), pp. 383–398.
[4] Tiwari, S., Tiwari, G.N. and Al-Helal, I.M., 2016. Development and recent trends in greenhouse dryer: A review. *Renewable and Sustainable Energy Reviews*, *65*, pp. 1048–1064.
[5] Müller, J., Reisinger, G., Kisgeci, J., Kotta, E., Tesic, M. and Mühlbauer, W., 1989. Development of a greenhouse-type solar dryer for medicinal plants and herbs. *Solar & Wind Technology*, *6*(5), pp. 523–530.
[6] Koyuncu, T., 2006. An investigation on the performance improvement of greenhouse-type agricultural dryers. *Renewable Energy*, *31*(7), pp. 1055–1071.
[7] Saini, V., Tiwari, S. and Tiwari, G.N., 2017. Environ economic analysis of various types of photovoltaic technologies integrated with greenhouse solar drying system. *Journal of Cleaner Production*, *156*, pp. 30–40.
[8] Sahu, T.K., Gupta, V. and Singh, A.K., 2016. A review on solar drying techniques and solar greenhouse dryer. *IOSR Journal of Mechanical and Civil Engineering (IOSR-JMCE)*, *13*(3), pp. 31–37.
[9] Patil, R. and Gawande, R., 2016. A review on solar tunnel greenhouse drying system. *Renewable and Sustainable Energy Reviews*, *56*, pp. 196–214.
[10] Kaewkiew, J., Nabnean, S. and Janjai, S., 2012. Experimental investigation of the performance of a large-scale greenhouse type solar dryer for drying chilli in Thailand. *Procedia Engineering*, *32*, pp. 433–439.
[11] Daliran, A., Taki, M., Marzban, A., Rahnama, M. and Farhadi, R., 2023. Experimental evaluation and modeling the mass and temperature of dried mint in greenhouse solar dryer; Application of machine learning method. *Case Studies in Thermal Engineering*, *47*, p. 103048.
[12] Çerçi, K.N. and Daş, M., 2019. Modeling of heat transfer coefficient in solar greenhouse type drying systems. *Sustainability*, *11*(18), p. 5127.
[13] Arunkumar, P.M., Balaji, N., Madhankumar, S. and Mohanraj, T., 2024. Prediction of red chilli drying performance in solar dryer with natural energy

storage element using machine learning models. *Journal of Energy Storage*, *101*, p. 113825.

[14] Abdelkader, T.K., Sayed, H.A., Refai, M., Ali, M.M., Zhang, Y., Wan, Q., Khalifa, I., Fan, Q., Wang, Y. and Abdelhamid, M.A., 2024. Machine learning, mathematical modeling and 4E (energy, exergy, environmental, and economic) analysis of an indirect solar dryer for drying sweet potato. *Renewable Energy*, *227*, p. 120535.

[15] Kushwah, A., Kumar, A. and Kumar, S., 2024. Predictive modelling with machine learning of garlic clove for novel designed solar drying system. *Solar Energy*, *284*, p. 113070.

[16] Cetina-Quiñones, A.J., Santamaria-Bonfil, G., Medina-Esquivel, R.A. and Bassam, A., 2023. Techno-economic analysis of an indirect solar dryer with thermal energy storage: An approach with machine learning algorithms for decision making. *Thermal Science and Engineering Progress*, *45*, p. 102131.

[17] Hu, G. and You, F., 2022. Renewable energy-powered semi-closed greenhouse for sustainable crop production using model predictive control and machine learning for energy management. *Renewable and Sustainable Energy Reviews*, *168*, p. 112790.

[18] Kushwah, A., Gaur, M.K., Kumar, A. and Singh, P., 2022. Application of ANN and prediction of drying behavior of mushroom drying in side hybrid greenhouse solar dryer: An experimental validation. *Journal of Thermal Engineering*, *8*(2), pp. 221–234.

[19] Paramasivam, P., Dzida, M., Osman, S.M., Le, D.T.N., Cao, D.N., Truong, T.H. and Tran, V.D., 2024. Development of advanced machine learning for prognostic analysis of drying parameters for banana slices using indirect solar dryer. *Case Studies in Thermal Engineering*, *60*, p. 104743.

[20] de Jesús Rubio, J., Garcia, E., Ochoa, G., Elias, I., Cruz, D.R., Balcazar, R., Lopez, J. and Novoa, J.F., 2019. Unscented Kalman filter for learning of a solar dryer and a greenhouse. *Journal of Intelligent & Fuzzy Systems*, *37*(5), pp. 6731–6741.

[21] Hao, W., Zhang, H., Liu, S. and Lai, Y., 2021. Design and prediction method of dual working medium solar energy drying system. *Applied Thermal Engineering*, *195*, p. 117153.

[22] Tagnamas, Z., Idlimam, A. and Lamharrar, A., 2023. Predictive models of beetroot solar drying process through machine learning algorithms. *Renewable Energy*, *219*, p. 119522.

[23] Venkateswaran, D. and Cho, Y., 2024. Efficient solar power generation forecasting for greenhouses: A hybrid deep learning approach. *Alexandria Engineering Journal*, *91*, pp. 222–236.

[24] Sangeetha, T., Thangadurai, K., Shankar, M., Nirmala, F.J. and Kumar, M.S., 2024, June. Development of solar based dryer for grapes drying using IoT and machine learning. In *AIP Conference Proceedings* (Vol. 3122, No. 1). AIP Publishing.

[25] Gorjian, S., Calise, F., Kant, K., Ahamed, M.S., Copertaro, B., Najafi, G., Zhang, X., Aghaei, M. and Shamshiri, R.R., 2021. A review on opportunities for implementation of solar energy technologies in agricultural greenhouses. *Journal of Cleaner Production*, *285*, p. 124807.

[26] Shankar, A., 2023. Efficient data interpretation and artificial intelligence enabled IoT based smart sensing system. *Artificial Intelligence Review*, *56*(12), pp. 15053–15077, https://doi.org/10.1007/s10462-023-10519-y.
[27] Molu, R.J.J., Tripathi, B., Mbasso, W.F., Naoussi, S.R.D., Bajaj, M., Wira, P. and Misak, S., 2024. Advancing short-term solar irradiance forecasting accuracy through a hybrid deep learning approach with Bayesian optimization. *Results in Engineering*, *23*, p. 102461, https://doi.org/10.1016/j.rineng.2024.102461.
[28] Tajjour, S., Chandel, S.S., Alotaibi, M.A., Malik, H., Márquez, F.P.G. and Afthanorhan, A., 2023. Short-term solar irradiance forecasting using deep learning techniques: A comprehensive case study. *IEEE Access*, *11*, 119851–119861, http://dx.doi.org/10.1109/ACCESS.2023.3325292.
[29] Sharma, S.K., Mallick, A., Singh, D.B. and Tiwari, G.N., 2022. Experimental study of solar energy-based water purifier of single-slope type by incorporating a number of similar evacuated tubular collectors. *Environmental Science and Pollution Research*, *29*(5), pp. 6837–6856, http://dx.doi.org/10.1007/s11356-021-16123-3.
[30] Amoussou, I., Paddy, E.Y., Agajie, T.F., Ibrahim, F.S., Agajie, E.F., Nsanyuy, W.B. and Mohammadi, S.A.D., 2024. Enhancing residential energy access with optimized stand-alone hybrid solar-diesel-battery systems in Buea, Cameroon. *Scientific Reports*, *14*(1), p.15543, https://doi.org/10.1038/s41598-024-66582-0.
[31] Singh, A.R., Kumar, R.S., Bajaj, M., Khadse, C.B. and Zaitsev, I., 2024. Machine learning-based energy management and power forecasting in grid-connected microgrids with multiple distributed energy sources. *Scientific Reports*, *14*(1), p. 19207, https://doi.org/10.1038/s41598-024-70336-3.
[32] Jamil, M., He, N., Wei, Z., Gupta, M.K. and Khan, A.M., 2023. A novel quantifiable approach of estimating energy consumption, carbon emissions and cost factors in manufacturing of bearing steel based on triple bottom-line approach. *Sustainable Materials and Technologies*, *36*, p. e00593, http://dx.doi.org/10.1016/j.susmat.2023.e00593.
[33] Chakrabarty, D., Patil, P. and Sharma, K., 2017. Experimental study on fluid flow and heat transfer from a rectangular prism approaching the wall of a wind tunnel. *Heat Transfer: Asian Research*, *46*(7), pp. 793–814, https://doi.org/10.1002/htj.21244.
[34] Bhandari, P., Padalia, D., Ranakoti, L., Khargotra, R., András, K. and Singh, T., 2023. Thermo-hydraulic investigation of open micro prism pin fin heat sink having varying prism sides. *Alexandria Engineering Journal*, *69*, pp. 457–468, https://doi.org/10.1016/j.aej.2023.02.016.
[35] Rekioua, D., Mokrani, Z., Kakouche, K., Rekioua, T., Oubelaid, A., Logerais, P.O. and Ghoneim, S.S., 2023. Optimization and intelligent power management control for an autonomous hybrid wind turbine photovoltaic diesel generator with batteries. *Scientific Reports*, *13*(1), p. 21830, http://dx.doi.org/10.1038/s41598-023-49067-4.
[36] Sathish, T., Saravanan, R., Jeyanthi, C.E., Sabariraj, R.V., Hemachandu, P., Ağbulut, Ü. and Soudagar, M.E.M., 2023. PCM embedded square channel receiver for effective operation of Net Zero Emission/Energy PDC building heating system. *Sustainable Energy Technologies and Assessments*, *60*, p. 103519, http://dx.doi.org/10.1016/j.seta.2023.103519.

[37] Kanouni, B., Badoud, A.E., Mekhilef, S., Bajaj, M. and Zaitsev, I., 2024. Advanced efficient energy management strategy based on state machine control for multi-sources PV-PEMFC-batteries system. *Scientific Reports*, *14*(1), p. 7996, https://doi.org/10.1038/s41598-024-58785-2.

[38] Almutairi, A., Albagami, N., Almesned, S., Alrumayh, O. and Malik, H., 2023. Electric vehicle load estimation at home and workplace in Saudi Arabia for grid planners and policy makers. *Sustainability*, *15*(22), p. 15878, http://dx.doi.org/10.3390/su152215878.

[39] Badri, Y.I., Sudabattula, S.K., Hussain, I., Malik, H. and Márquez, F.P.G., 2023. Ultracapacitor as selectable energy buffer in electric vehicle application. *Journal of Energy Storage*, *73*, p. 109200, http://dx.doi.org/10.1016/j.est.2023.109200.

[40] Tiwari, S., Swaminathan, M. and Harender, S.D.B., 2022. Performance enhancement of the photovoltaic system with different cooling methods. *Environmental Science and Pollution Research*, *29*(30), pp. 45107–45130, https://link.springer.com/article/10.1007/s11356-022-20330-x.

[41] Pimoli, D.S., Tomar, P., Kumar, A., Dwivedi, G. and Kumar Shukla, A., 2025. Experimental investigation of friction factor and heat transfer enhancement using circular perforated twisted tape inserts in heat exchangers. *International Journal of Energy for a Clean Environment*, https://doi.org/10.1615/InterJEnerCleanEnv.2024053616.

[42] Tomar, P., Chauhan, P., Kumar, A. and Kumar Shukla, A., 2024. Numerical investigation of thermal performance enhancement of solar reservoir using flash cycle. *Journal of Advanced Research in Fluid Mechanics and Thermal Sciences*, *123*(1), 197–221, https://doi.org/10.37934/arfmts.123.1.197221.

[43] Verma, T., Singh, T.S., Rajak, U., Nashine, P., Dwivedi, G. and Kumar, A., 2024. *Clean Energy: Technology, Advances and Applications*. Boca Raton, FL: CRC Press, https://doi.org/10.1201/9781003521341.

[44] Kumar, A., Singh, V.P., Meena, C.S. and Dutt, N., 2023. *Thermal Energy Systems: Design, Computational Techniques and Applications*. Boca Raton, FL: CRC Press, https://doi.org/10.1201/9781003395768.

[45] Kumar, A., Agrawal, P.S. and Singla, Y.K., 2025. *Sustainable Development With Carbon Capture, Utilization and Storage Technology*. Switzerland: Springer Nature.

Chapter 3

Application of IoT in Solar Dryer Design

Neetu Singh, V. N. Bartaria, Ajeet Kumar Soni, Prashant Baredar, and Satish Pal Singh Rajput

3.1 INTRODUCTION

The traditional methods of agricultural goods preservation are still being used by many of their producers and the people who are involved in mass storage of them. These methods include drying of goods, keeping refrigerated goods, curing many food products, preparing pickles, and storing in cans and bottles. Among all these and many more such methods, the drying of food and agricultural products is the age-old and most favorable method [1]. Drying chiefly involves moisture removal from the food products by retarding the chemical reactions in them. It kills the pathogens that lead to decay in the foods. As the moisture is needed for the growth of most of the micro-organisms, the reduction in moisture content up to an acceptable level by dehydrating foods will reduce the possibility of its decay. This provides the opportunity for food producers to preserve their food products while ensuring the good quality of the products. A good market value of the food products can also be ensured by their proper preservation [2].

Drying has always been a practice for the preservation of agricultural products. This aims to retain the quality of the product and its flavor. The commonly used practices in food drying are making use of electricity, coal, fossil fuels, or direct solar rays. These significantly alter the quality of food products or adversely affect the environment [3]. As a promising alternative, solar energy is being used as a green energy in food drying. Solar drying is gaining significant popularity in recent years due to an increasing focus on the production of quality agricultural goods using solar energy and other environmental concerns.

A solar dryer, which harnesses renewable energy from the sun and wind, is designed to boost natural air circulation within a building. As a key example of solar-assisted passive ventilation systems, which are included in the structure of the system [4, 5].

DOI: 10.1201/9781003634737-3

3.2 SOLAR ENERGY

Electricity is the primary need for the smooth conduction of most of the routine personal and commercial work of everyone. Major electricity production takes place by combustion of fossil fuels like oil, gas, and coal. It is another fact that the source of one-third of the global greenhouse emissions is due to the electricity generation from these fuels. To provide a better livelihood and raise the living standard of the people in society, a clean and green source of energy is necessary. To reduce global emissions, green energy sources such as solar energy can be harnessed in many applications where electricity is needed or thermal energy can be used [5, 6].

The sun provides an abundant amount of energy on the surface of the earth. The global horizontal irradiation varies at different locations on the earth. The average solar radiation intensity received on the surface for India is about 200 MW/km^2 [7]. The global horizontal irradiation as the sum of the amount of direct and diffused solar radiation, as shown in Figure 3.1, hits a horizontal surface on the earth [8].

The use of solar energy helps in reducing greenhouse gas emissions, in addition to fostering development in the rural areas, where the conventional source of energy as electricity is either not available or is produced from alternative means. Use of solar energy also helps in improving energy security, and therefore emphasis on its usage is continuously increasing [8]. India is contributing to the research and development of technologies based

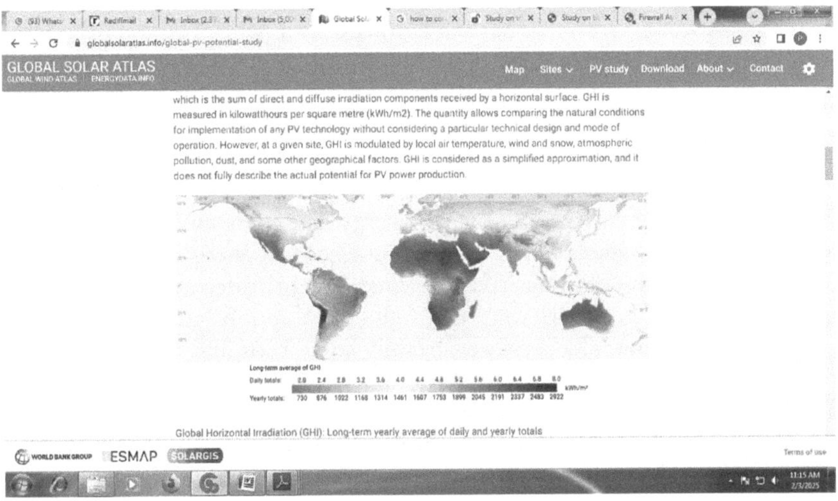

Figure 3.1 Global horizontal irradiation [8].

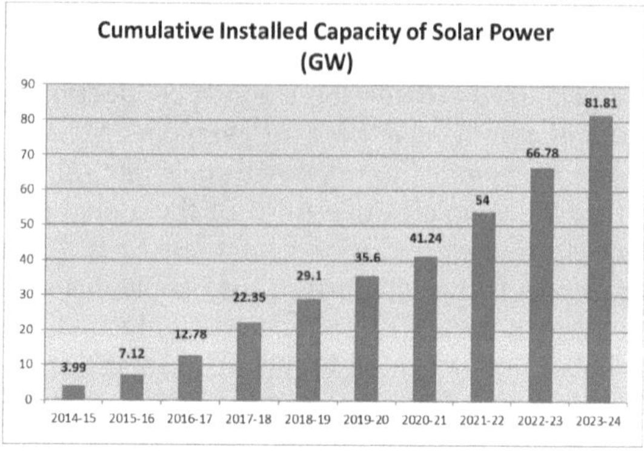

Figure 3.2 Cumulative installed capacity of solar power (GW).

Source: Ministry of New and Renewable Energy (MNRE)

on various non-conventional energy sources such as biomass, wind, ocean energy, geothermal energy, and solar energy [9, 10].

In recent years, solar energy has been emerging as a dominant force, with the cumulative installed capacity as of March 31, 2024, from solar energy as 81.81 GW. Figure 3.2 shows the growth in cumulative installed capacity of solar power (GW) in India. This is significantly higher than the cumulative installed capacity of 3.99 GW as of March 31, 2015, as per the Ministry of New and Renewable Energy (MNRE), Govt of India. This shows the significant and dramatic increase in the total installed capacity of solar energy in India, with a growth of 1850.38% during this period. Among the total installations of solar power, 81.88% are ground-mounted installations [11].

3.3 SOLAR DRYER

The drying of agricultural goods needs energy. The energy in drying will need a huge amount of electricity if done using electrical energy. This also leads to severe environmental hazards. The green energy alternative is the most attractive option as no conventional fuel consumption is required and no pollution is associated. There are certainly some pros and cons in the type of energy being used in dryers. Many researchers and designers of dryers are therefore focusing on the design and development of solar dryers with a hybrid mode. Recent focus of researchers and solar dryer designers is on the design and development of solar dryers that are capable of drying under controlled conditions while retaining the food quality [12].

The drying of food products requires moisture removal from them by the dry air. Air movement helps in carrying the moisture away from the food products [13]. Solar dryers utilize solar energy for moisture removal and air movement for the effective drying of agricultural products [14–16]. The design and development of various solar drying techniques has taken place from the development of sun drying methods over a long time. The focus on this design development was to make a closed system that can trap the solar radiation in the form of solar energy to increase the inner temperature for drying [17].

3.4 DESIGN PROCESS

At the initial phase of the dryer design process, the design parameters are determined based on the information of available dryer designs, study of the climatic data, and solar insolation at the particular location. The collector configuration is then decided carefully before the dimensions of the dryer are finalized. It is needed to analyze the performance of such dryer designs, and certain modifications are incorporated in the initial design on the basis of the performance evaluation [18, 19]. A typical design of solar flat plate collector suitable for tropical areas for the dryer is shown in Figure 3.3.

It essentially has a glass cover of 5 mm thickness, has an absorber surface of GI sheet having a thickness of about 0.6 mm, and is painted black to increase the heat absorption. Insulation with wood, especially plywood, as a good insulating material, is commonly used in solar collectors. Air gap between the wooden sheets is also provided to reduce heat loss [20].

3.5 SOLAR DRYER CLASSIFICATION

Solar dryers are based on clean and green energy technology using solar energy for drying food products. This is an advantageous technology as compared to open sun drying and other conventional drying methods. Solar dryers may be classified on many criteria, some of which are the drying temperature range, air circulation method, and incident solar radiation on the absorber. A greenhouse dryer is another type of dryer using greenhouse structure for the drying of agricultural goods. It has better drying efficiency and less loss on crops [21].

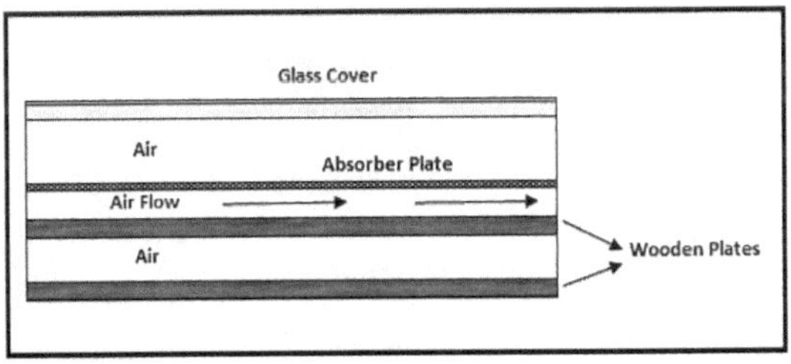

Figure 3.3 Solar energy collector for solar dryer.

The literature reveals that many researchers have been continuously exploring new design features. Numerous experimental and numerical studies show the advancements in design and performance analysis of the solar dryers. An empirical study conducted during winter and summer seasons shows the performance of greenhouse solar dryer for non-operating conditions. The results of the study show that the temperature range in summers is 7.6°C–13.4°C while for winters 6.3°C–13.2°C. The humidity range found is 26%–47% in summer and 23%–40% in winter. So, it is interesting to know that the trends are very similar [22]. Table 3.1 shows the general classification of solar dryers.

Table 3.1 Classification of Solar Dryers Table 3.1: Table shows the general classification of solar dryers.

Criteria	*Type*	*Advantages*	*Disadvantages*
Method of air circulation	Natural Convection Type	• Less cost • Less maintenance • No blower required and thus external source of energy	• Not suitable during non-sunny days • Possibility of deterioration in quality of product
	Forced Convection Type	• Better efficiency and rate of drying as compared to natural convection type	• Requires blower and thus external source of energy • More maintenance involved
Exposure of solar radiation to the goods	Direct Type	• Simple and cost-effective design • Having single unit for heat collection and drying	• Low drying rate • Large space required for the same quantity of drying as compared to indirect dryer
	Indirect Type	• High drying rate • Smaller space required for the same quantity of drying as compared to indirect dryer	• Complex in design • More cost
	Mixed Mode Type	• Higher rate of drying due to drying by hot air and direct solar radiation • Controlled drying conditions	• Higher cost involvement • Higher maintenance
	Hybrid Type	• Controlled drying conditions • Higher rate of drying • Day and night drying	• More expensive • Higher maintenance due to complex design structure

3.6 IoT-BASED SOLAR DRYER

Solar dryer design development is rapidly taking place with more interest among researchers and designers for making its design more effective and efficient. With the advent of modern technology and the desire for more control over the drying operation and control over drying quality, the design needs to use networking infrastructure [23]. With the use of the Internet of Things (IoT), the operating parameters and working conditions of the dryer can be continuously monitored for the effective use of the dryer through sensors, software, and internet.

The solar dryer can be connected to the internet for useful information interchange by using Radio Frequency Identification (RFID), Global Positioning System (GPS), laser scanners, and other connected devices [24]. The smart use of solar energy with modern display of all the controlling parameters and ease will make it more useful and applicable. This will also enable the user to have greater control over system monitoring [25].

3.7 PRINCIPLE OF WORKING

Conventional solar dryers can incorporate IoT technology by integrating information and sensor technologies. This allows for the electronic collection and storage of data pertaining to various operating parameters, such as temperature and humidity, within the dryer system. For this, the requisite sensors will be required for placing them at suitable and desired locations. For the operation of electrical and electronic components, electrical energy can be obtained from a PV module. The useful information collected will be provided to the IoT platform in the form of data. For the effective control of the drying process in the dryer, any minor change in the information received may then be used through the internet [25].

3.8 CONCEPTUAL SYSTEM CONFIGURATION

The use of renewable energy in daily applications can be enormously increased by making use of smart and modern internet-enabled technologies. This will enable the user to have more control over the system operation and its effective use. IoT solutions provide more user-friendly operation and control of processes throughout the world.

The IoT-based solar dryer system should conceptually have the appropriate configuration for serving its purpose, along with an application program:

- PV module
- Charge controller

- Battery of suitable size and capacity
- Thermocouples of appropriate type
- Relays
- Humidity sensors
- Flow sensors

The application program will display all the information received from the IoT platform for the set time interval. It may be collected at 1-minute intervals. The process of drying will then be monitored and controlled from a remote location. With the ease in such operation through remote control, the efficiency and effectiveness of the dryer will greatly improve [26, 27].

3.9 DESIGN MODULE

A solar dryer with IoT can be designed to get an improved version, which can reduce the spoilage of the agricultural products by effective control of the various parameters such as drying temperature, humidity, and air velocity. A design module of the solar dryer with IoT can be developed using the PV module, charge controller, battery, microcontroller with networking facility, various sensors for measuring temperature, humidity, and air flow. A typical arrangement is shown in Figure 3.4. This system will enable the remote user to get the gathered data from the sensors of the module for effective monitoring of the drying process. This will also let the user control the drying process in the desired way.

The sensors will have the ability to detect the temperature, humidity, and air velocity inside the dryer. The physical state of the agricultural product

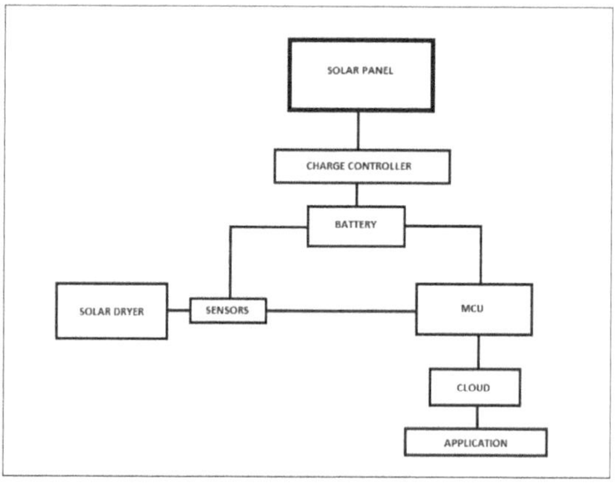

Figure 3.4 IoT-based solar dryer module.

kept inside the dryer will also be known from the data gathered by the sensors at a set time interval. The solar panel will generate electricity when it receives solar radiation on the panel. The battery will be charged from the output of the panels after regulating the charge in the charge controller. The battery will serve two important purposes in this module by providing energy to the blower and to the ECU. The drying parameters will be controlled by the ECU.

The drying conditions are controlled with the aid of an Arduino or Node MCU. Arduino board contains a microcontroller, power supplies, crystals, and female headers for interfacing with the peripherals. Various typical Arduino boards, such as ATmega8, ATmega168, ATmega328, ATmega1280, and ATmega2560, may be of use. Arduino programs are written in C or C++ and only two functions are defined to make a program, while other details are kept hidden from the user. Although the programming of Arduino is simpler and easier.

On the other hand, a Node MCU with sensors may be used. It is an open-source firmware and is a low-cost open-source IoT platform. It has firmware running on the ESP8266. Power from the battery is supplied to the Node MCU to recharge it continuously. Setting of temperature, humidity, and flow sensors is done before the programming is fed into the Node MCU. Connections are then established. This Node MCU with ESP-8266 microcontroller chip has greater memory and built-in Wi-Fi connectivity. It utilizes an internet server for sending the data received from various sensors to the application.

The dryer module with IoT can be monitored and operated remotely through the Wi-Fi internet connectivity using the NODE MCU, and the application is an Android app that allows you to limit your internet usage. Blynk was designed with the Internet of Things in mind. It has the ability to manage hardware remotely, display sensor data, record data, visualize it, and perform a wide range of other amazing things. With the Blynk app smartphone application can be created that will allow the user to interact with the Node MCU microcontroller, such as the Raspberry Pi.

For a typical IoT-based solar dryer, the system configuration may include a solar panel of 400–500W, charge controller of 12V-10A capacity, 12V battery, Node MCU with ESP-8266 microcontroller chip, K-type thermocouples of suitable temperature range, humidity sensor, and air flow velocity sensors. The output from the sensors is supplied to the Node MCU, which is further sent to the server. All the information sent to the server is obtained on the screen of the application for the user to have greater monitoring of the dryer performance. An interface will allow the control of operating parameters to improve performance of the solar dryer.

To analyze the sensitivity of this IoT-based solar dryer, the data as received from the temperature and humidity sensors may be compared with those gathered from measuring through the thermometer and humidity meter to calculate the percentage variation between them.

3.10 CONCLUSION

This chapter presents the design of an IoT-based solar dryer system for improving the drying operations through better monitoring and control of various operating parameters. The IoT-based solar dryer has improved monitoring and control through the application program displaying all the information of the system parameters as received and collected at the IoT platform. The usage of the application will make solar dryer more useful and easier in drying. The information and conceptual system configuration presented here will enable the researchers and designers to design a smart solar dryer for their specific application for a specific crop or agricultural product to be dried using the green energy application as solar energy.

A design module with system component details and a suitable specification is also provided. An IoT-based solar dryer system with greater monitoring capability and control of operating parameters from a remote location may thus be designed. System sensitivity may also be analyzed by calculating the percentage variation in the data gathered through the temperature and humidity sensors.

REFERENCES

[1] Rosa Rolle S (2006) *Postharvest management of fruit and vegetables in the Asia-Pacific region.* Tokyo, Japan: Asian Productivity Organization.

[2] Ghaly AE, MacDonald KN (2012) An effective passive solar dryer for thin layer drying of poultry manure. Am J Eng Appl Sci 5(2):136–150.

[3] Abdullahi Y, Momoh M, Garba MM, Musa M (2013) Design and construction of an adjustable and collapsible natural convection solar food dryer. Int J Comput Eng Res 3(6):1–8.

[4] Bartaria VN, Rajput SPS (2013) Energy-efficient technology of air distribution in air conditioning. TERI Inf Digest Energy Environ 12(2):183–188.

[5] Rajput KS, Bartaria VN (2024) Experimental study of vertical air temperature distribution and ventilation with energy saving using solar chimney. Lecture Notes in Civil Engineering, https://doi.org/10.1016/j.matpr.2022.04.914.

[6] Rao KV, Bartaria VN, Bhave P (2019) Experimental study on ventilation through solar chimney. Int J Civil Eng Technol 10(12):545–549.

[7] Hemaiswarya S, Raja R, Carvalho IS, Ravikumar R, Zambare V, Barh D (2012) An Indian scenario on renewable and sustainable energy sources with emphasis on algae. Appl Microbiol Biotechnol 96:1125–1135, https://doi. org/10.1007/s00253-012-4487-0.

[8] Global Solar Atlas (2025) Global horizontal Irradiation obtained from the "Global Solar Atlas 2.0", a free, web-based application [Accessed 25 January 2025], https://globalsolaratlas.info/map.

[9] Pal S, Mukhopadhyay T (2023) An overview on pre and post COVID 19 electrical energy requirement, consumption and generation in India and present situation. J Math Anal Appl 3:20–27, https://doi.org/10.55529/jecnam.33.20.27.

[10] Shu T, Papageorgiou DJ, Harper MR, Rajagopalan S, Rudnick I, Botterud A (2022) From coal to variable renewables: Impact of flexible electric vehicle charging on the future Indian electricity sector. Energy 269:126465, https://doi.org/10.1016/j.energy.2022.126465.

[11] Chakraborty K, Mukherjee S, Paul S (2021) Indian energy scenario and smart grid development. In: *Advances in Clean Energy Technologies: Select Proceedings of ICET 2020*. Singapore: Springer, 171–176, https://doi.org/10.1007/978-981-16-0235-1_14.

[12] Ahmad A, Prakash O, Kumar A, Chatterjee R, Sharma S, Kumar V, Kulshreshtha K, Li C, Eldin EM (2022) A comprehensive state-of-the-art review on the recent developments in green house drying. Energies 15:9493, https://doi.org/10.3390/en15249493.

[13] Fernandes L, Fernandes JR, Tavares PB (2022) Design of a friendly solar food dryer for domestic over-production. Solar 2:495–508.

[14] Jain D, Tewari P (2015) Performance of indirect through pass natural convective solar crop dryer with phase change thermal energy storage. Renew Energy 80:244–250.

[15] Singh S, Kumar S (2012) Testing method for thermal performance based rating of various solar dryer designs. Sol Energy 86:87–98.

[16] Almuhanna EA (2011) Utilization of a solar greenhouse as a solar dryer for drying dates under the climatic conditions of the eastern province of Saudi Arabia. J Agric Sci 4:237–246.

[17] Rajkumar P, Kulanthaisami S, Raghavan GSV, Gariépy Y, Orsat V (2007) Drying kinetics of tomato slices in vacuum assisted solar and open sun drying methods. Dry Technol 25:1349–1357.

[18] Vlachos NA, Karapantsios TD, Balouktsis AI, Chassapis D (2002) Design and testing of a new solar tray dryer. Dry Technol 20(6):1243–1271.

[19] Pangavhane DR, Sawhney RL (2002) Review of research and development work on solar driers for grape drying. Energy Convers Manag 43(1):45–61.

[20] Sukhatme SP (2000) *Solar Energy, Principles of Thermal Conduction and Storage*. New Delhi, India.

[21] Prakash O, Kumar A (2014) Solar green house drying: A review. Renew Sustain Energy Rev 29:905–910, https://doi.org/10.1016/j.rser.2013.08.084.

[22] Mondru Madhava MM, Sivala Kumar SK, Rao DB, Smith DD, Kumar HV (2017) Performance evaluation of photo-voltaic ventilated hybrid green house dryer under no-load condition. Agric Eng Int CIGRJ 19:93–101.

[23] Wadaskar A, Chouriya N, Alam T, Yadav S, Ninave V, Shaha R (2021) IoT based solar hybrid dryer. Int J Adv Eng Manag 3(8):1680–1684.

[24] Harischandrakar S, Raul A, Mahajan A, Bakliwal J, Pandit N (2023) Design and development of IOT based smart solar dryer. Int J All Res Educ Sci Methods (IJARESM) 11(9):127–132.

[25] Patil SM, Vijayalashmi RTM (2021) IoT based solar energy monitoring system. Mater Today Proc 1574–1579, https://doi.org/10.1016/j.matpr.2021.07.364.
[26] Reyes A, Cubillos F, Mahn A, Vásquez J (2014) Control system in a hybrid solar dryer. In: *2014 5th International Renewable Energy Congress (IREC)*, 5–9, https://doi.org/10.1109/IREC.2014.6826936.
[27] Cui Y, Liu W, Rani P, Alrasheedi M (2021) Internet of Things (IoT) adoption barriers for the circular economy using Pythagorean fuzzy SWARA-CoCoSo decision-making approach in the manufacturing sector. Technol Forecast Soc Change 171:120951.

Chapter 4

Computer-Based Interface System for Online Testing of Parameters for Solar Dryers

Shan Mahto and Krishna Varma

4.1 INTRODUCTION

Agriculture is the most important sector in India. More than 60% of the population of India is directly dependent on agriculture and related activities for livelihood. Agricultural activities include cultivation, harvesting, processing, storage, and many more. One such activity is drying or dehydration. Drying refers to the process of removing excess moisture from a product to its safe or desired level of moisture content [1–3]. The drying process can be applied to various agricultural products, including fruits such as mango, grapes, banana, blueberry, and figs; vegetables such as garlic, ginger, onion, potato, mint, moringa leaves, coriander, fenugreek, curry leaves, and bee pollen; medicinal herbs such as tulsi, neem, and ashwagandha. Major benefits of drying include increased shelf life, protection against rotting of produce, reduced transportation weight, improved product quality, and better market value. A majority of population depends upon direct sunlight, which is also termed as open sun drying, to carry out the drying process for their agricultural produce. Although open sun drying is an easy process and requires little investment, it is prone to risk of contamination and abrupt weather changes [4, 5]. Other limitations of utilizing open sun drying include lightweight products such as bee pollen, leaves, and herbs being blown away by the wind, over-drying of products, as well as lack of quality product [6–9].

Solar dryers utilize solar thermal energy for effectively drying agricultural products. Indirect solar dryers overcome most of the challenges of open sun drying effectively. As compared to open sun drying, solar dryers generate higher temperatures and lower relative humidity of product, provide controlled drying environment, and provide high-quality product [10–12].

The indirect solar dryers developed at research institutes such as MGIRI Wardha, India, are being sent for field trials in various rural areas of India, such as villages of Vijaywada (Andhra Pradesh), Leh (Ladakh), Makhla (Maharashtra), villages of Rajasthan, Gujarat, and Tamil Nadu. The travel and accommodation expenditures of the institute officers for conducting field trials at these villages spread all across the country are a major concern.

DOI: 10.1201/9781003634737-4

The Internet of Things (IoT) is a dynamic system where a variety of devices with dissimilar communication capabilities engage in synchronization and exchange of data over the internet [3]. Direct updation of data is achieved without any human intervention [4]. An IoT microcontroller unit (MCU) consists of user interface, hardware devices including sensors, actuators and processors, backend software, and connectivity. The integration of IoT is necessary for the migration of traditional industries to smart industries. Many researches such as [5–8] have discussed various applications of IoT for monitoring of data utilizing precise digital sensors for measurement of various parameters such as temperature and humidity.

Researches such as [9–11] show the utilization of ThingSpeak IoT platform for the analysis of real-time data from sensors in the form of graphs, tables, charts, etc.

The challenges involved in conducting on-site field trials include travel and accommodation expenses of the officers. Moreover, the field trials involve manual monitoring, which requires opening of the solar dryer box door for taking measurements [13–15]. This leads to inaccuracies in measurements. Along with this, the data collection and data evaluation processes become time-consuming due to the absence of any kind of mechanism for real-time monitoring and visualization. This research aims to overcome these challenges of conducting on-site field trials by providing a computer interface system for conducting the field trials completely online and monitoring critical parameters such as temperature and humidity for solar dryers. DHT22 sensors are interfaced with an ESP8266 Wi-fi Node MCU board [16–22]. These sensors collect temperature and humidity data from the solar dryer located at the installation site and are directly transmitted to the computers at the research institute utilizing Internet of Things. Data analysis and verification of these parameters can be done directly within the premises of the institute. This eliminates the cost of travel and accommodation of officers and also saves time and effort [23–32].

4.2 OBJECTIVES

The primary objectives of this research are outlined as follows:

1. To provide an online interface for monitoring parameters like temperature and humidity for solar dryers.
2. To reduce travel and accommodation expenditures of officers and save time by providing a computer interface for conducting field trials online.
3. To provide more precise and accurate plotting of temperature and humidity graphs in real-time at continuous intervals.
4. To provide fast and accurate data evaluation and data visualization.

5. To evaluate the agricultural products for moisture content and analyze parameter values graphically over the internet.
6. To eliminate the efficiency loss that occurred during on-site measurement of temperature and humidity.

4.3 MATERIALS AND METHODS

This study was conducted as a part of a project implemented at Mahatma Gandhi Institute for Rural Industrialization (abbreviated as MGIRI), a research institute of national importance under the Ministry of Micro, Small, and Medium Enterprises (abbreviated as MSME), located in the city of Wardha, Maharashtra, India. The research begins with the interfacing of DHT22(Digital Humidity and Temperature) sensors with the ESP8266 NodeMCU board. The connection diagram (Figure 4.2) is shown. The programming and optimization of the code for the Node MCU board is done on the Arduino Integrated Development Environment (IDE). Once the code was optimized, consistency of readings was checked for all DHT22 sensors. All the sensors provided consistent readings. The sensors were placed in different sections of the dryer. A total of six DHT22 sensors were placed in different sections of the solar dryer. The first sensor was placed inside the flat plate collector, the second sensor was placed at the passage, the third in the base of the Dryer, the fourth sensor was placed over the lower tray, the fifth sensor at the upper tray, and the sixth sensor was placed at the exhaust vent. Moringa leaves (*Moringa oleifera)* were placed in the trays of the dryer for testing of the system. The NodeMCU was provided with Wi-Fi using a Jio-Fi Module, enabling data transfer directly to the computer. We have provided two options for data analysis and visualization; the first one is ThingSpeak IoT platform, and the second one is a self-developed interface using the Streamlit application. The process flow is given in Figure 4.1.

The diagram shows six DHT22 sensors connected to an ESP8266 NodeMCU, which processes temperature and humidity data. The power supply is a 12V DC battery, which is powered by a solar panel. The Wi-Fi module enables real-time data transmission to the cloud via an Application Programming Interface (API). The data is then analyzed on a computer for monitoring and visualization.

4.4 HARDWARE ASPECTS

- **ESP8266 NodeMCU Microcontroller Board:** The ESP8266 NodeMCU is a compact, cost-effective microcontroller board that provides full Wi-Fi connectivity along with a complete TCP/IP stack. It offers multiple interfaces such as UART, SPI, I2C, and a wide range of GPIO pins,

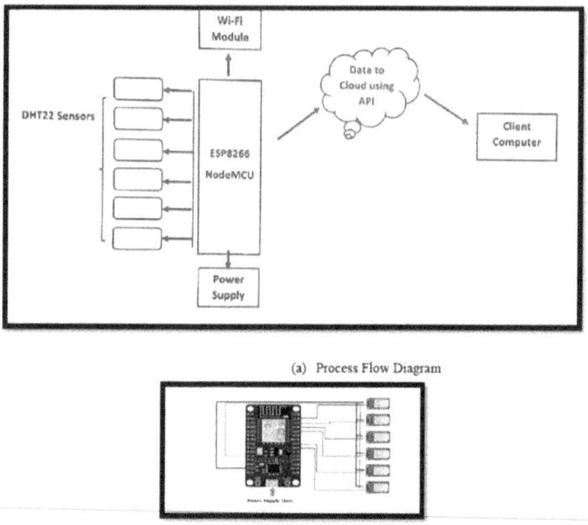

(a) Process Flow Diagram

(b) Interfacing DHT22 sensors with ESP8266 NodeMCU board

Figure 4.1 Process flow diagrams and interfacing.

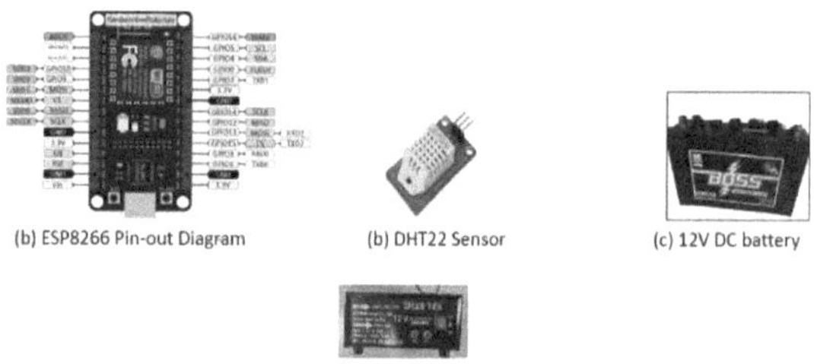

(b) ESP8266 Pin-out Diagram (b) DHT22 Sensor (c) 12V DC battery

(d) AC to DC converter

Figure 4.2 Different components of interface systems.

making it ideal for interfacing with various sensors and peripherals. The board comes equipped with a built-in voltage regulator, a CP2102 USB-to-UART converter for easy programming via the Arduino IDE and operates at a low voltage (2.5–3.3V). Its robust design and integrated features enable reliable and continuous data collection in IoT applications. Figure 4.2 shows a Pin-out description of NodeMCU board.

- **DHT22 Sensor:** The DHT22 sensor is a low-cost digital device used to measure ambient temperature and humidity with high accuracy. Operating on 3–5V with minimal current consumption, it utilizes a capacitive humidity sensor and a thermistor to generate digital outputs without the need for an analogue input. With an accuracy of ±0.5°C for temperature and 2–5% for humidity, the DHT22 is well-suited for applications requiring precise environmental monitoring, despite its relatively low sampling rate of once every 2 seconds.

- **12V DC Battery (Power Supply):** The 12V DC battery provides a stable and reliable power source for the system. Measured in ampere-hours (Ah), the battery ensures that the connected components receive consistent power over extended periods. Its design, based on lead plates that create a potential difference between the positive and negative terminals, makes it suitable for both stationary and mobile applications. Proper maintenance—such as regulated charging and protection from extreme temperatures—is essential to sustain performance and prolong the battery's lifespan.
- **12V DC to AC Converter (Inverter):** The 12V DC to AC converter, or inverter, transforms the battery's direct current (DC) into alternating current (AC), which is often required by certain devices and components. This conversion process involves an oscillator circuit that generates a high-frequency AC waveform, which is then boosted by a transformer to achieve the desired output voltage and provide electrical isolation. In this project, the inverter is crucial for delivering the correct voltage to the NodeMCU board and other peripherals, ensuring stable operation even under fluctuating power conditions.
- **Wi-Fi Module:** While the ESP8266 NodeMCU board includes integrated Wi-Fi functionality, a dedicated Wi-Fi module may also be employed to enhance connectivity and network performance. This module is responsible for establishing and maintaining a robust wireless connection using preset credentials (SSID and password), ensuring that sensor data is reliably transmitted to online platforms. It supports real-time data exchange and is optimized for high data throughput and strong signal integrity, which is critical for continuous monitoring and remote access in IoT applications.

4.5 SOFTWARE ASPECTS

4.5.1 ThingSpeak IoT Platform

- Arduino IDE: An open-source development environment for programming Arduino microcontrollers, offering a user-friendly code editor with syntax highlighting, auto-completion, and a serial monitor for real-time debugging.
- ThingSpeak™ IoT Platform: A cloud-based platform for collecting, visualizing, and analyzing sensor data in real-time, featuring MATLAB® integration for advanced analytics and APIs for seamless connectivity with other services.

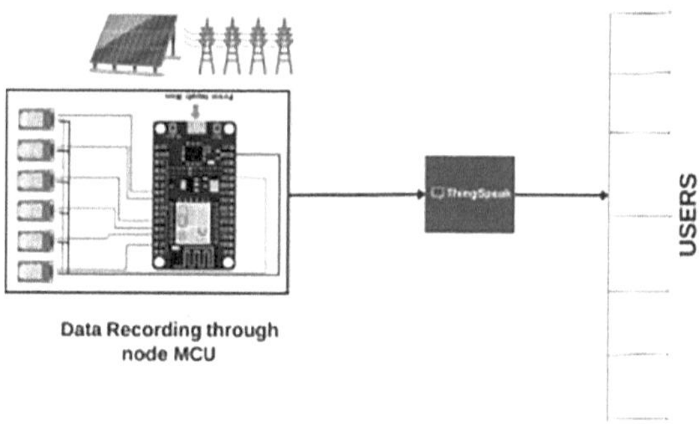

Figure 4.3 Process flow for ThingSpeak.

Working: The following points (See Figure 4.3) explain the working of the ThingSpeak IoT platform:

1. **Creating an Account in ThingSpeak:** To start using ThingSpeak, users first navigate to the ThingSpeak website and sign up with their credentials. During the registration process, a secure password is set up, completing the account creation. This step is essential to gain access to the platform's full suite of IoT tools and services.
2. **Setting Up a Channel:** After account creation, a new channel is established by entering the required fields, such as the channel name, description, and other pertinent details. Upon saving, the platform automatically generates a unique Channel ID. This identifier is crucial for distinguishing and managing the data stream associated with that channel.
3. **API Key Management:** ThingSpeak automatically issues API keys for each channel, which are vital for data transmission and security. The Write API Key is used for uploading sensor data to the channel, while the Read API Key enables data retrieval and visualization. Maintaining the security of these API keys is critical, as they ensure that only authorized users have access to your data.
4. **Integration with Arduino IDE/NodeMCU:** The generated API keys are embedded within the NodeMCU code programmed via the Arduino IDE. This integration enables the microcontroller to transmit sensor data directly to the ThingSpeak channel in real-time. Consequently, the system supports continuous data collection, monitoring, and analysis, thereby enhancing the effectiveness of IoT-based applications.

4.5.2 Streamlit-Based Self-Developed Application

- Arduino IDE: Arduino Integrated Development Environment is used to program the microcontroller Node MCU to collect sensor data and give various other commands to the microcontroller (Figure 4.4).

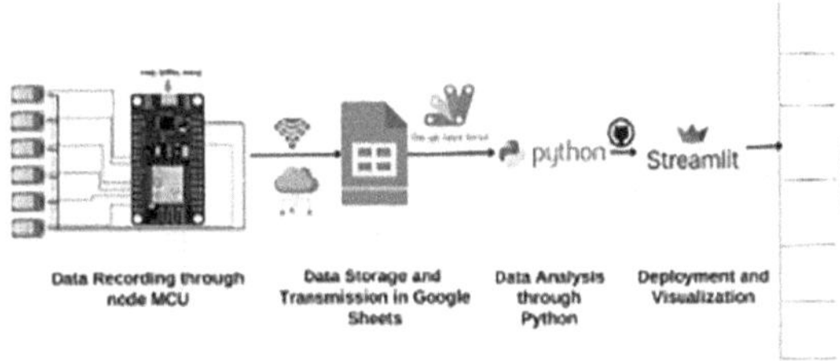

Figure 4.4 Process flow for Streamlit application.

- Google Sheets: Google Sheets is a cloud-based spreadsheet tool that allows real-time data storage and sharing. Google Sheets is used as a storage and transmission interface as it provides easily accessible and structured format for further analysis.
- App Scripts: Google Apps Script is a JavaScript-based cloud scripting platform that automates tasks and integrates Google services. Here, it is used to automatically store the incoming sensor data in proper rows and columns.
- Python: Python is a versatile programming language widely used for data analysis, automation, and backend processing. Here, Python is used to extract the sensor data, analyze, and visualize the data.
- GitHub: GitHub is a version control and collaboration platform used for code management and project documentation. GitHub is used to make a repository of files to host them on Streamlit, track changes in the project, and empower collaboration.
- Streamlit: Streamlit is an open-source Python framework for creating interactive and user-friendly web applications. Streamlit is used to provide a real-time dashboard to analyze and visualize data.

4.5.3 Working

1. **Sensor Initialization and Data Collection:** The microcontroller is programmed to initialize all the sensors integrated into the system. Wi-Fi credentials (SSID and password) are preset to ensure a seamless connection to the available network near the dryer. A deliberate 15-minute delay is implemented between sensor readings to ensure that data is captured regularly and accurately. During each cycle, sensor data is captured and stored in designated variables, preparing it for subsequent transmission.
2. **Data Transmission and Storage:** Once the sensor data is collected, it is transmitted over the internet to Google Sheets. A Google Apps Script embedded within the spreadsheet automatically organizes the incoming data into predefined columns—such as Date, Time, Temperature S1, Humidity S1—and appends each new data entry as a new row. This

automated process ensures that the data is systematically stored and readily available for further analysis.

3. **Data Integration and Analysis:** A user-friendly interface is developed using Streamlit, which features two main pages: the Overview and the Visualize page. The Overview page displays key metrics such as each dryer's latest monitoring date, average temperature, and average humidity. Meanwhile, the Visualize page allows users to select a specific dryer, date, and parameter to generate various types of plots—including bar, line, distribution, and scatter plots—providing comprehensive visual insights into the data.
4. **Deployment and Accessibility:** The entire application, comprising the microcontroller code and the Python interface, is maintained within a GitHub repository, ensuring efficient version control and collaboration. Once deployed, the application is accessible from anywhere, enabling users to monitor and analyze the data in real-time, regardless of their location.

4.6 RESULTS AND DISCUSSIONS

The testing and verification of the system were done, and data visualizations for every section of the solar dryer where the sensors were placed were available. Two different options for data visualization in the form of ThingSpeak IoT platform and Streamlit application are provided as per the requirement of the Institute. ThingSpeak IoT platform is a subscription-based platform, whereas the Streamlit application is a free-of-cost application completely programmed by us. Hence, cost-related options are available with the Institute. The visualizations and graphs of real-time temperature and humidity data are shown in Figures 4.5, 4.6, and 4.7.

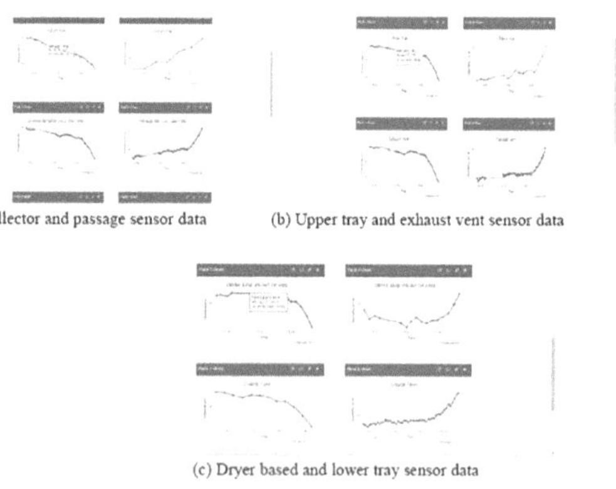

(a) Collector and passage sensor data

(b) Upper tray and exhaust vent sensor data

(c) Dryer based and lower tray sensor data

Figure 4.5 ThingSpeak IoT platform visualizations.

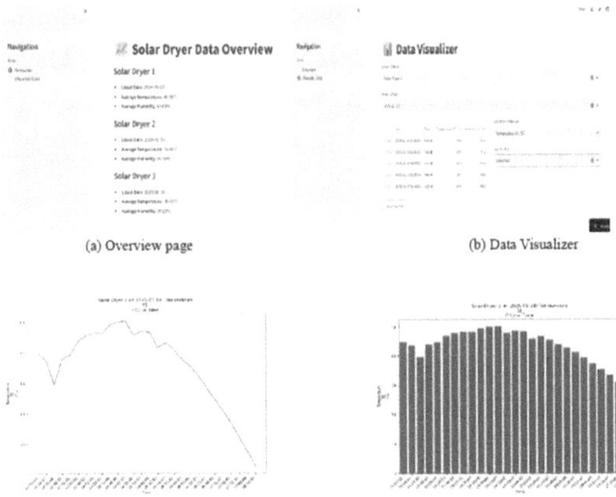

(a) Overview page (b) Data Visualizer
(c) Line Chart (d) Bar Chart

Figure 4.6 Streamlit application visualizations.

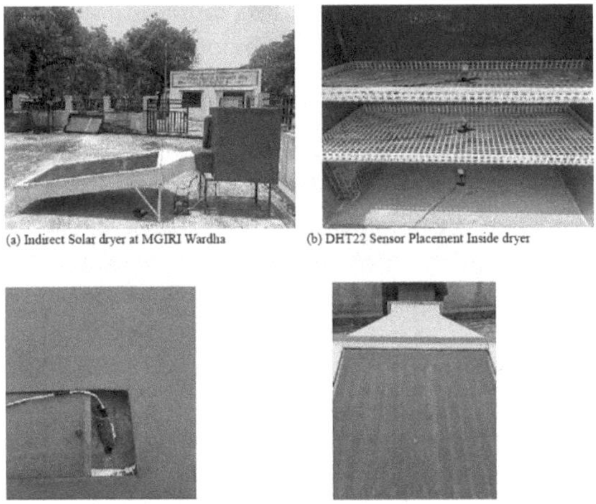
(a) Indirect Solar dryer at MGIRI Wardha (b) DHT22 Sensor Placement Inside dryer
(c) DHT22 Sensor Placement in exhaust vent (d) DHT22 Sensor Placement Inside collector

Figure 4.7 Project implementation and solar dryer.

4.7 CONCLUSIONS AND FUTURE SCOPE

The computer-based interface system offers several advantages, including reduced travel and accommodation costs by enabling remote monitoring and data access. It minimizes manual effort and human error through automation, ensuring accurate and efficient data collection, processing, and analysis. The system increases efficiency of the data measurement process as there is no need for opening the solar dryer box for taking temperature and humidity measurements manually as digital temperature and humidity sensors are directly installed in various sections of the solar dryer. These sensors fetch real-time parameters such as temperature and humidity without interfering with the equilibrium of the solar dryer. This ensures accurate measurements of temperature and humidity. The system is scalable and adaptable, allowing it to be implemented across various agricultural and industrial applications where measurements of humidity and temperature are necessary. The data collection and analysis can now be done for a large number of solar dryers located in various areas of the country. This eliminates the process of on-site field trials, making the field trial process fast, efficient, and scalable. Two systems were developed to provide cost-related options to the research institute. The first option is ThingSpeak IoT platform, which has a subscription fee that varies based on the number of visualizations or graphs required by the institute. The second option is a Streamlit application developed by us, which is completely free to use for the institute. The real-time data helps in identifying any system malfunctions almost instantly. This ensures proper drying of the agricultural product and helps in maintaining optimum quality of the dried product. A good quality of dried agricultural product has a good shelf life. Sales of these dried products provide an additional income stream to the farmers, improving their financial stability. This can help in improving the quality of life for the farmers.

The future scope of this computer-based system includes integration with AI and ML for predictive analysis and automation, enabling smarter decision-making and enhanced efficiency. The available additional slots in the IC board provide flexibility for incorporating more sensors and actuators, expanding the system's functionality. Further advancements can include the incorporation of control mechanisms to regulate and maintain desired drying parameters, ensuring optimal conditions within the dryer. Additionally, advanced data analytics can be leveraged to further optimize drying efficiency, improving performance and resource utilization. Beyond the current application, the system can be expanded to other agricultural and industrial drying processes, making it a versatile and scalable solution for diverse use cases.

REFERENCES

[1] Srivastava, A., Sharma, A., & Kumar, A. (2025). Performance evaluation of indirect solar drying system for potato slices: Comparative analysis with open-sun drying method. Solar Energy, 285, 113114. https://doi.org/10.1016/j.solener.2024.113114.

[2] Purohit, P., Kumar, A., & Kandpal, T. C. (2006). Solar drying vs. open sun drying: A framework for financial evaluation. Solar Energy, 80(12), 1568–1579. https://doi.org/10.1016/j.solener.2005.12.009.

[3] Kuo, Y. F., Chen, L. C., & Tseng, M. L. (2018). Constructing an effective management system for the agriculture industry based on IoT. IEEE Access, 6, 44464–44474.

[4] Faber, C., Müller, J., & Ortwein, A. (2020). Optimization of solar drying systems: A review. Renewable Energy, 160, 1378–1398.

[5] Parihar, Y. S. (2019). *Internet of things and Nodemcu A review of use of Nodemcu ESP8266 in IoT products*. https://www.researchgate.net/publication/337656615_Internet_of_Things_and_Nodemcu_A_review_of_use_of_Nodemcu_ESP8266_in_IoT_products.

[6] Macheso, P. S., Manda, T. D., Meela, A. G., Mlatho, J. S., Taulo, G. T., & Phiri, J. C. (2021, December). Industrial temperature monitor based on NodeMCU ESP8266, MQTT and Node-RED. In *2021 3rd International Conference on Advances in Computing, Communication Control and Networking (ICAC3N)* (pp. 740–743). IEEE. https://doi.org/10.1109/ICAC3N53548.2021.9725469.

[7] Hasan, M. W. (2023). Building an IoT temperature and humidity forecasting model based on long short-term memory (LSTM) with improved whale optimization algorithm. *Memories-Materials, Devices, Circuits and Systems*, 6, 100086. https://doi.org/10.1016/j.memori.2023.100086.

[8] Jabbar, W. A., Ting, T. M., Hamidun, M. F. I., Kamarudin, A. H. C., Wu, W., Sultan, J., & Ali, M. A. (2024). Development of LoRaWAN-based IoT system for water quality monitoring in rural areas. *Expert Systems With Applications*, 242, 122862. https://doi.org/10.1016/j.eswa.2023.122862.

[9] Chandrasekaran, G., Kumar, N. S., Chokkalingam, A., Gowrishankar, V., Priyadarshi, N., & Khan, B. (2025). IoT enabled smart solar water heater system using real time ThingSpeak IoT platform. *IET Renewable Power Generation*, 19(1), e12760. https://doi.org/10.1049/rpg2.12760.

[10] Parida, D., Behera, A., Naik, J. K., Pattanaik, S., & Nanda, R. S. (2019, May). Real-time environment monitoring system using ESP8266 and ThingSpeak on internet of things platform. In *2019 International Conference on Intelligent Computing and Control Systems (ICCS)* (pp. 225–229). IEEE. https://doi.org/10.1109/ICCS45141.2019.9065451.

[11] Razali, M. A. A., Kassim, M., Sulaiman, N. A., & Saaidin, S. (2020, June). A ThingSpeak IoT on real time room condition monitoring system. In *2020 IEEE International Conference on Automatic Control and Intelligent Systems (I2CACIS)* (pp. 206–211). IEEE. https://doi.org/10.1109/I2CACIS49202.2020.9140127.

[12] Shankar, A. (2023). Efficient data interpretation and artificial intelligence enabled IoT based smart sensing system. *Artificial Intelligence Review*, 56(12), 15053–15077. https://doi.org/10.1007/s10462-023-10519-y.

[13] Molu, R. J. J., Tripathi, B., Mbasso, W. F., Naoussi, S. R. D., Bajaj, M., Wira, P., & Misak, S. (2024). Advancing short-term solar irradiance forecasting accuracy through a hybrid deep learning approach with Bayesian optimization. *Results in Engineering*, 23, 102461. https://doi.org/10.1016/j.rineng.2024.102461.

[14] Tajjour, S., Chandel, S. S., Alotaibi, M. A., Malik, H., Márquez, F. P. G., & Afthanorhan, A. (2023). Short-term solar irradiance forecasting using deep learning techniques: a comprehensive case study. *IEEE Access*, 11, 119851–119861. http://dx.doi.org/10.1109/ACCESS.2023.3325292.

[15] Sharma, S. K., Mallick, A., Singh, D. B., & Tiwari, G. N. (2022). Experimental study of solar energy-based water purifier of single-slope type by incorporating a number of similar evacuated tubular collectors. *Environmental Science and Pollution Research*, 29(5), 6837–6856. http://dx.doi.org/10.1007/s11356-021-16123-3.

[16] Amoussou, I., Paddy, E. Y., Agajie, T. F., Ibrahim, F. S., Agajie, E. F., Nsanyuy, W. B., Bajaj, M., & Mohammadi, S. A. D. (2024). Enhancing residential energy access with optimized stand-alone hybrid solar-diesel-battery systems in Buea, Cameroon. *Scientific Reports*, 14(1), 15543. https://doi.org/10.1038/s41598-024-66582-0.

[17] Singh, A. R., Kumar, R. S., Bajaj, M., Khadse, C. B., & Zaitsev, I. (2024). Machine learning-based energy management and power forecasting in grid-connected microgrids with multiple distributed energy sources. *Scientific Reports*, 14(1), 19207. https://doi.org/10.1038/s41598-024-70336-3.

[18] Jamil, M., He, N., Wei, Z., Gupta, M. K., & Khan, A. M. (2023). A novel quantifiable approach of estimating energy consumption, carbon emissions and cost factors in manufacturing of bearing steel based on triple bottom-line approach. *Sustainable Materials and Technologies*, 36, e00593. http://dx.doi.org/10.1016/j.susmat.2023.e00593.

[19] Chakrabarty, D., Patil, P., & Sharma, K. (2017). Experimental study on fluid flow and heat transfer from a rectangular prism approaching the wall of a wind tunnel. *Heat Transfer: Asian Research*, 46(7), 793–814. https://doi.org/10.1002/htj.21244.

[20] Bhandari, P., Padalia, D., Ranakoti, L., Khargotra, R., András, K., & Singh, T. (2023). Thermo-hydraulic investigation of open micro prism pin fin heat sink having varying prism sides. *Alexandria Engineering Journal*, 69, 457–468. https://doi.org/10.1016/j.aej.2023.02.016.

[21] Rekioua, D., Mokrani, Z., Kakouche, K., Rekioua, T., Oubelaid, A., Logerais, P. O., Ali, E., Bajaj, M., Berhanu, M., & Ghoneim, S. S. (2023). Optimization and intelligent power management control for an autonomous hybrid wind turbine photovoltaic diesel generator with batteries. *Scientific Reports*, 13(1), 21830. http://dx.doi.org/10.1038/s41598-023-49067-4.

[22] Sathish, T., Saravanan, R., Jeyanthi, C. E., Sabariraj, R. V., Hemachandu, P., Ağbulut, Ü., & Soudagar, M. E. M. (2023). PCM embedded square channel receiver for effective operation of Net Zero Emission/Energy PDC building heating system. *Sustainable Energy Technologies and Assessments*, 60, 103519. http://dx.doi.org/10.1016/j.seta.2023.103519.

[23] Kanouni, B., Badoud, A. E., Mekhilef, S., Bajaj, M., & Zaitsev, I. (2024). Advanced efficient energy management strategy based on state machine control for multi-sources PV-PEMFC-batteries system. *Scientific Reports*, 14(1), 7996. https://doi.org/10.1038/s41598-024-58785-2.

[24] Almutairi, A., Albagami, N., Almesned, S., Alrumayh, O., & Malik, H. (2023). Electric vehicle load estimation at home and workplace in Saudi Arabia for grid planners and policy makers. *Sustainability*, 15(22), 15878. http://dx.doi.org/10.3390/su152215878.

[25] Badri, Y. I., Sudabattula, S. K., Hussain, I., Malik, H., & Márquez, F. P. G. (2023). Ultracapacitor as selectable energy buffer in electric vehicle application. *Journal of Energy Storage*, 73, 109200. http://dx.doi.org/10.1016/j.est.2023.109200.

[26] Tiwari, S., Swaminathan, M., Harender, & Singh, D. B. (2022). Performance enhancement of the photovoltaic system with different cooling methods. *Environmental Science and Pollution Research*, 29(30), 45107–45130. https://link.springer.com/article/10.1007/s11356-022-20330-x.

[27] Kundu, A., Kumar, A., Awasthi, M. K., Dwivedi, G., & Shukla, A. K. (2025). An explorative study of earth-air heat exchanger for sustainable building applications. *International Journal of Energy for a Clean Environment*, 26(3), 1–22. https://doi.org/10.1615/InterJEnerCleanEnv.2024052513.

[28] Razaviyn, Z., Heidari, M., Thangavel, S., Verma, V., Kumar, A., & Yadav, A. K. (2025). Marine energy harvesting from fluid flow via vortex induced vibrations. *International Journal of Thermofluids*, 25, 101015. https://doi.org/10.1016/j.ijft.2024.101015.

[29] Pimoli, D. S., Tomar, P., Kumar, A., Dwivedi, G., & Shukla, A. (2025). Experimental investigation of friction factor and heat transfer enhancement using circular perforated twisted tape inserts in heat exchangers. *International Journal of Energy for a Clean Environment*, 26(3). https://doi.org/10.1615/InterJEnerCleanEnv.2024053616.

[30] Mokhtari, M., Jamei, S., Razaviyan, Z., Heidari, M., Thangavel, S., & Kumar, A. (2024). Numerical and hydrodynamic analysis of a container vessel due to the change of collision wave angle at different speeds in regular wave. *Journal of Mechanical Engineering and Sciences*, 10247–10260. https://doi.org/10.15282/jmes.18.4.2024.3.0809.

[31] Sharma, N., Kumar, A., Dutt, N., & Kumar, A. (2022). Experimental study and performance optimization of a ribbed cooling passage by Taguchi method. *Journal of Mechanical Science and Technology*, 209, 118250.

[32] Saini, S., Tomar, P., Kumar, A., & Dutt, N. (in press). Numerical investigation of heat transfer enhancement and flow performance of a tubular heat exchanger with surface modifications. *Heat Transfer*.

Chapter 5

Artificial Intelligence-Based Predictive Modeling for Accurate Solar Power Generation

The Comparison between SVR, ANN, and RFR Approaches

Prof. Dileep Kumar Mohanachandran, Dr. Uppin Chandrashekhar, Dr. Kalpana Venugopal, Dr. Gopala T, and Achyutha Prasad N

5.1 INTRODUCTION

As a feasible renewable energy source, solar power generation has an important position in the global reduction of carbon emissions and shift toward renewable energy solutions (Ibrahim et al. 2012). By 2023, global installed solar capacities are going to reach more than 1,000 GW, increasing the demand for accurate prediction of solar power generation (Uhlir and Schröder 2007). Solar energy is highly variable by its nature and weather patterns, seasons, and atmospheric disturbances, and thus has made accurate predictions of power generation a great challenge (Behera and Nayak 2020). To optimize energy distribution, to minimize grid instability and facilitate an efficient integration of solar power into the electricity market, accurate forecasting models are needed (Kim and Kim 2016).

Classic methods of solar forecasting rely on statistical and physics-based methods that rely on historical weather data, solar radiation models, and empirical computation. The techniques are used extensively but tend to fail when dealing with intricacies and nonlinearity of solar energy variability, producing suboptimal predictive accuracy (Zhang et al. 2017). To avoid this, artificial intelligence (AI) and machine learning (ML) strategies became popular as an alternative since they can handle big data and automatically pick complex patterns and provide an adaptive solution for forecasts (Quiles et al. 2020). Traditional methods cannot handle different meteorological parameters like solar irradiance, temperature, humidity, wind speed, and cloud cover, while AI-based models are more powerful, more efficient, and able to deal with them (Monís et al. 2020).

With the incorporation of AI in solar power forecasting, different sophisticated models like support vector regression (SVR), artificial neural networks

DOI: 10.1201/9781003634737-5

(ANN), and random forest regressor (RFR) have been created. Each of these models has distinct strengths: in terms of nonlinear relationships and small datasets, SVR is especially good; ANN is very useful for finding complex feature relationships by way of deep learning; and RFR is an ensemble method to provide better accuracy and interpretability (King, Sansom, and Comley 2020). These models use historical solar power generation data together with real-time meteorological data to enhance the prediction accuracy. Moreover, AI methods can also adapt to changing environmental conditions and thus make more accurate and dynamic forecasting for solar energy systems (Kiesecker et al. 2020).

The objective of this research is to develop and compare these AI-driven models, viz. SVR, ANN, and RFR for accurate solar power forecasting. Our main tasks are to make available both meteorological and historical solar power data; preprocess and train and optimize each model with hyperparameter tuning; measure our performance in terms of root mean square error (RMSE), mean absolute error (MAE), and R^2 Score; and finally to compare the uniqueness and flaws of each method. This research compares these models in depth and aims to identify the best AI-driven forecasting method for solar power generation. Moreover, these results will be useful for the larger renewable energy forecasting community regarding the practical application of models, model interpretability, and possible improvements, such as hybrid AI methods for higher accuracy and reliability.

5.2 LITERATURE REVIEW

The growing integration of solar energy into the world's electricity grid has made solar power forecasting an issue of key interest (Madeshwaren et al. 2024). Good insight into whether energy is required in the short or long term is important for grid stability and energy management, as well as cost savings. Many modes have been developed, such as simple statistical methods up to AI-based methods (Subramanian et al. 2023). Historically, there has been a move from traditional time series analysis to current ML and deep learning methods, which have higher accuracy and flexibility (Alkahtani, Aldhyani, and Alsubari 2023).

Persistence models, physical models, and statistical models (e.g., autoregressive integrated moving average, ARIMA; multiple linear regression, MLR) represent conventional forecasting techniques. The first group of these techniques forecasts the power generation by using past weather and solar irradiation data (Barrera et al. 2020). While this benchmark is set by them in terms of prediction, they are found to have a pronounced weakness when addressing highly nonlinear and dynamic weather conditions. However, physical models such as the clear sky model and numerical weather prediction (NWP) are computationally expensive and are less flexible in how they consider local weather trends such as fog (International Energy Agency 2023).

Due to their ability to work with nonlinearities and derive meaningful patterns from big data, AI forecasting models provide a better alternative than the usual methods (Gala et al. 2014). Some of the most popular AI models for solar power forecasting are SVR, ANN, and RFR. Kernel-based learning method SVR is the best fit for small- to medium-sized datasets and capable of finding complex relationships among meteorological variables (Quiles et al. 2020). The brain-inspired model ANN uses several levels of connected neurons to understand the complex dependencies in the solar power generation data (Subramanian et al. 2023). Combining advantages of statistical and neural learning, RFR, an ensemble learning method based on decision trees, is applicable for feature importance analysis and for overfitting robustness (Alkahtani et al. 2023).

In recent years, there have been many other studies of AI for solar forecasting. The comparison of several ML models, namely SVR, ANN, and RFR for short-term solar power forecasting was carried out by Barrera et al. (2020), and therefore their research reiterates that SVR showed better performance than ANN and RFR in terms of R^2 score and RMSE values. According to Gala et al. (2014), ANN is good at functioning under highly volatile weather conditions because of its deep learning structure. Mirjalili et al. (2023) also studied the trade-off between accuracy and interpretability, finding that RFR is more informative of the feature importance but perhaps not as accurate as ANN.

While the state of the art of AI-based solar forecasting has improved over the past years, there are still some research gaps. The main challenge, on the other hand, is the need for hybrid models that fuse the power of several algorithms to enhance predictive performance. Research suggests that combining ANN with SVR takes advantage of ANN's functional capabilities and SVR's generalization power (Mirjalili et al. 2023). Moreover, uncertainty quantification of AI models is underdeveloped, yet this is essential in improving the reliability of AI models in real-world applications (Nguyen et al. 2025). Additionally, the forecast system may be made real time, with adaptive learning mechanisms that update the AI models with new data, arriving constantly, so it can also improve forecast accuracy in a dynamic environment (Lee et al. 2018).

Finally, with increased accuracy and flexibility over conventional approaches, AI-based approaches have greatly improved the forecasting of solar power. However, issues such as algorithm interpretability, computational complexity, and hybridization of algorithms need further investigation. This research builds on previous studies by comparing SVR, ANN, and RFR for predicting solar power generation and choosing the best method for its use in renewable energy management.

Table 5.1 Description of Input Features Used in the Study

Feature	*Description*	*Units*
Solar Irradiance	Amount of solar energy received per unit area	W/m^2
Temperature	Ambient temperature in Celsius	°C
Humidity	Atmospheric moisture level (%)	%
Wind Speed	Speed of wind affecting solar panel cooling	m/s
Cloud Cover	Percentage of sky covered with clouds	%
Pressure	Atmospheric pressure in hPa	hPa

5.3 METHODOLOGY

5.3.1 Data Collection and Pre-Processing

This research utilizes data from weather stations, satellite imagery, and historic records of solar power generation. However, all these sources have detailed meteorological parameters that are critical to solar power generation (Simon Haykin (McMaster University, Hamilton, Ontario) 2005). For this research, we employ the features that are essential: solar irradiance, temperature, humidity, wind speed, cloud cover, and atmospheric pressure, as seen in Table 5.1.

Missing values are treated using linear interpolation and mean imputation methods in order to ensure data quality (Adugna, Ramu, and Haldorai 2024). Furthermore, all features are scaled (min-max scaling into a common range), to increase model stability (Hanandeh, Ardah, and Abu-Farsakh 2020). To facilitate models to detect time-dependent trends in solar power generation, feature engineering has been implemented by creating lag variables and rolling means (Simon Haykin (McMaster University, Hamilton, Ontario) 2005).

5.4 MACHINE LEARNING MODELS

5.4.1 Support Vector Regression (SVR)

In particular, SVR is optimized for capturing intricate, nonlinear patterns inherent in solar power forecasting. It achieves this through the utilization of kernel functions, which efficiently map input features to a higher dimension, and enables it to uncover what turns out to be very subtle patterns in meteorological data. However, there may be a problem with its performance since hyperparameter selection of C, epsilon (ε), and gamma (γ) can lead to underfitting or overfitting, if they are inappropriate. Though SVR is highly accurate, its computational cost increases with larger datasets and is not efficient for real-time applications.

5.4.2 Artificial Neural Networks (ANN)

It presents an approach for modeling intricate correlations among meteorological variables and solar power output using an extensive set of meteorological data. Their multi-layered structure enables ANNs to learn deep features in input data and improve predictive performance. The same optimizers like Adam and SGD used in our model make backpropagation faster, resulting in faster convergence, and ReLU activation function helps us take training to the next level. Dropout layers plus regularization are used to increase generalization because ANNs are very susceptible to overfitting (particularly for small datasets). One weakness of ANNs, relative to simple models including RFR, is that they are not easily interpretable.

5.4.3 Random Forest Regressor (RFR)

It's an ensemble approach, that is, it constructs several decision trees and averages their predictions to be more stable and less overfitting. Feature importance analysis is especially valuable with RFR, helping us find the most important meteorological parameters for solar forecasting. RFR is the easiest to use for large data and also most robust when dealing with missing values at a low computational complexity but with an increasing complexity as the number of trees increases. RFR is extremely interpretable but is relatively poor at extrapolating patterns beyond its training range, and so is less suitable for extreme weather conditions.

5.5 EXPERIMENTAL SETUP AND RESULTS

5.5.1 Dataset Description

The dataset used in this research includes historical weather station records containing meteorological data, and the corresponding historical solar output record from weather stations, satellite data, and energy monitoring systems. Crucial weather variables that have a huge impact on solar power generation are present in the dataset, such as solar irradiance, temperature, humidity, wind speed, cloud cover, and atmospheric pressure. These are measured at hourly time intervals and thus capture short-term fluctuations in environmental conditions. Table 5.2 shows key descriptive statistics of the dataset in the form of mean, standard deviation, min, and max, for each variable.

The variation in feature values shows how dynamic weather conditions affect solar power generation. Factors such as solar irradiance and cloud cover experience notable changes, which directly influence the efficiency of solar panels.

Table 5.2 Summary Statistics of Dataset Features

Feature	*Mean*	*Std Dev*	*Min*	*Max*
Solar Irradiance	312.14	10.52	41.62	488.75
Temperature	761.56	87.29	10.62	362.11
Humidity	590.96	62.11	9.09	650.67
Wind Speed	486.95	72.27	9.17	225.54
Cloud Cover	141.69	6.96	15.21	362.93
Pressure	141.68	97.14	26.24	429.73

Table 5.3 Optimal Hyperparameter Settings for Each Model

Model	*Hyperparameters*
SVR	Kernel: RBF, C: 100, Epsilon: 0.1, Gamma: Scale
ANN	Layers: [64, 32, 16], Activation: ReLU, Optimizer: Adam
RFR	Estimators: 100, Max Depth: None, Min Samples Split: 2

5.6 MODEL TRAINING AND HYPERPARAMETER TUNING

To improve model performance, hyperparameter tuning is performed for SVR, ANN, and RFR using GridSearchCV to systematically optimize parameters. The chosen optimal hyperparameters are outlined in Table 5.3.

For SVR, the RBF kernel is selected because it performs exceptionally well with nonlinear relationships. The regularization parameter C is set to 100, which strikes a good balance between model complexity and generalization. The epsilon (ε) margin is set at 0.1, indicating the acceptable range for prediction errors. In the case of ANN, a three-layer architecture with [64, 32, 16] neurons per layer is implemented, utilizing ReLU activation to boost learning efficiency. The Adam optimizer is chosen for its adaptive learning features, which enhance convergence speed. For RFR, 100 estimators (trees) are utilized to ensure robustness while avoiding excessive computational demands. The maximum depth is not restricted, allowing the trees to develop naturally, and the minimum samples per split is set to 2 to maintain complexity.

5.7 COMPARISON OF MODEL PERFORMANCE

The models are assessed using root mean square error (RMSE), mean absolute error (MAE), R^2 Score, and computational time (s). Table 5.4 displays the comparative results.

Table 5.4 Performance Comparison of SVR, ANN, and RFR

Model	*RMSE*	*MAE*	*R^2 Score*	*Computational Time (s)*
SVR	28.24	17.86	0.852	4.77
ANN	41.41	19.81	0.743	4.85
RFR	17.99	6.16	0.716	4.14

From the results, we see that SVR has the best predictive accuracy out of the models with an R^2 of 0.852. However, RFR, with 17.99 RMSE and 6.16 MAE, had lowest error. While the advanced deep learning capabilities of ANN are demonstrated, it had a high RMSE of 41.41 and training time of 4.85 seconds, suggesting overfitting issues. Another important factor to consider is computational efficiency—RFR was the fastest, taking 4.14 seconds, and more easily suitable for real-time applications, while SVR and ANN needed a bit longer, 4.77 seconds and 4.85 seconds, respectively. Findings highlight the tradeoffs between prediction accuracy, error minimization, and computation time among the various models.

5.8 RESULTS ANALYSIS AND DISCUSSION

In Figure 5.1, we compare the actual and predicted solar power output using SVR, ANN, and RFR. The colored lines, dotted, show the predictions of all different models, and the black dashed line shows actual values. According to the graph, the SVR matches the actual trend so well, but there are slight problems in dealing with the extreme amplitude, on account of SVR's sensitivity to settings of hyperparameters. The correlation between SVR prediction and actual values is high with R^2 score of 0.852; however, it is important to note that it has RMSE of 28.24 W and MAE of 17.86 W, which suggests some prediction errors. Although ANN is able to fit complex relationships, there is a significant deviation of its predictions near sharp peaks and troughs, indicative of generalization issues and perhaps overfitting. ANN has the lowest R^2 score of 0.743 with RMSE of 41.41 W and MAE of 19.81 W and the largest errors and the weakest capability to track fluctuations. However, RFR provides the closest match to the actual values with minimal deviations and demonstrates good predictive performance as well as robustness to noise; however, it tends to smooth rapid fluctuations, which may result in missing short-term variability.

5.9 CONCLUSIONS

The aim of this study is to investigate the application of SVR, ANN, and RFR on forecasting solar power from historical meteorological data. Their findings established that one model was suited for a particular task, but not for

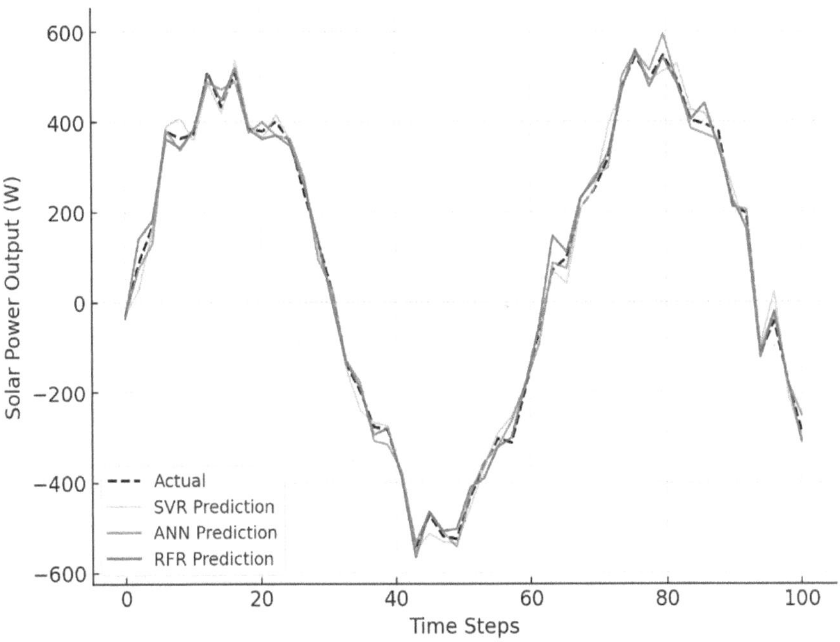

Figure 5.1 Actual versus predicted solar power output for all models.

another. Of the three models, SVR drew the highest R^2 score (0.852), meaning it was strongly correlated with actual solar power output. However, it has fairly good error values (RMSE = 28.24 W, MAE = 17.86 W) and is also quite costly computationally (4.77s training time). Since the accuracy must be high, SVR could be the right choice, if sufficient computational resources exist. However, RFR also achieved the lowest RMSE (17.99 W) and MAE (6.16 W), and therefore the best error reduction. The quick training time (4.14s) and its robustness and efficiency also made it very tempting. However, RFR had a tendency to reduce short-term randomness, and hence might not be as good in capturing very high or low changes of solar power output as SVR. In comparison to other algorithms, ANN proved to be the most flexible and suitable for nonlinear relationship modeling, while displaying the highest RMSE (41.41 W) and MAE (19.81 W) and longest training time (4.85 sec). Therefore, this means that ANN does support complex relationships, and the same also applies to complex feature generation, such as the genetic algorithms, but there is still room for improvement with ANN—scaling data, regularization (such as early stopping) and perhaps more data will help with generalization and prevent overfitting. One important finding of this study is that RFR is the most balanced among the models and offers the best compromise between prediction accuracy and computational cost, and it is particularly well suited for use in practical solar power forecasting. For instance, SVR works best when accuracy is the requisite point but does not restrict from a computational point of view; on the other hand, ANN needs additional adjustment to enhance its performance. Further research could then investigate hybrid models that merge SVR and RFR and include deep learning methods, such as long short-term memory (LSTM) networks for time series forecasting.

REFERENCES

Adugna, Teshome Debushe, Arulmurugan Ramu, and Anandakumar Haldorai. 2024. "A Review of Pattern Recognition and Machine Learning." *Journal of Machine and Computing* 4(1):210–220.

Alkahtani, Hasan, Theyazn H. H. Aldhyani, and Saleh Nagi Alsubari. 2023. "Application of Artificial Intelligence Model Solar Radiation Prediction for Renewable Energy Systems." *Sustainability (Switzerland)* 15(8). https://doi.org/10.3390/su15086973.

Barrera, Jose Manuel, Alejandro Reina, Alejandro Maté, and Juan Carlos Trujillo. 2020. "Solar Energy Prediction Model Based on Artificial Neural Networks and Open Data." *Sustainability (Switzerland)* 12(17). https://doi.org/10.3390/SU12176915.

Behera, Manoja Kumar, and Niranjan Nayak. 2020. "A Comparative Study on Short-Term PV Power Forecasting Using Decomposition Based Optimized Extreme Learning Machine Algorithm." *Engineering Science and Technology, an International Journal* 23(1):156–167. https://doi.org/10.1016/j.jestch.2019.03.006.

Gala, Y., A. Fernández, J. Dorronsoro, M. García, and C. Rodríguez. 2014. "Machine Learning Prediction of Global Photovoltaic Energy in Spain." *Renewable Energy and Power Quality Journal* 1(12):605–610. https://doi.org/10.24084/repqj12.423.

Hanandeh, Shadi, Allam Ardah, and Murad Abu-Farsakh. 2020. "Using Artificial Neural Network and Genetics Algorithm to Estimate the Resilient Modulus for Stabilized Subgrade and Propose New Empirical Formula." *Transportation Geotechnics* 24(February):100358. https://doi.org/10.1016/j.trgeo.2020.100358.

Ibrahim, S., I. Daut, Y. M. Irwan, M. Irwanto, N. Gomesh, and Z. Farhana. 2012. "Linear Regression Model in Estimating Solar Radiation in Perlis." *Energy Procedia* 18:1402–1412. https://doi.org/10.1016/j.egypro.2012.05.156.

International Energy Agency. 2023. "World Energy Outlook 2023|Enhanced Reader," 23–28.

Kiesecker, Joseph, Sharon Baruch-Mordo, Mike Heiner, Dhaval Negandhi, James Oakleaf, Christina Kennedy, and Pareexit Chauhan. 2020. "Renewable Energy and Land Use in India: A Vision to Facilitate Sustainable Development." *Sustainability (Switzerland)* 12(1). https://doi.org/10.3390/su12010281.

Kim, Sungil, and Heeyoung Kim. 2016. "A New Metric of Absolute Percentage Error for Intermittent Demand Forecasts." *International Journal of Forecasting* 32(3):669–679. https://doi.org/10.1016/j.ijforecast.2015.12.003.

King, Peter, Christopher Sansom, and Paul Comley. 2020. "Photogrammetry for Concentrating Solar Collector Form Measurement, Validated Using a Coordinate Measuring Machine." *Sustainability (Switzerland)* 12(1):1–20. https://doi.org/10.3390/su12010196.

Lee, Woonghee, Keonwoo Kim, Junsep Park, Jinhee Kim, and Younghoon Kim. 2018. "Forecasting Solar Power Using Long-Short Term Memory and Convolutional Neural Networks." *IEEE Access* 6(c):73068–73080. https://doi.org/10.1109/ACCESS.2018.2883330.

Madeshwaren, Vairavel, Kathiresan Govindaraju, Pragya Pandey, Suresh Thangavel, and Rajpradeesh Thangaraj. 2024. "Predictive Modeling for Solar Desalination Using Artificial Neural Network Techniques: A Review." *Global Nest Journal* 26(2). https://doi.org/10.30955/gnj.005481.

Mirjalili, Mohammad Amin, Alireza Aslani, Rahim Zahedi, and Mohammad Soleimani. 2023. "A Comparative Study of Machine Learning and Deep Learning Methods for Energy Balance Prediction in a Hybrid Building-Renewable Energy System." *Sustainable Energy Research* 10(1). https://doi.org/10.1186/s40807-023-00078-9.

Monís, Julián Ignacio, Rafael López-Luque, Juan Reca, and Juan Martínez. 2020. "Multistage Bounded Evolutionary Algorithm to Optimize the Design of Sustainable Photovoltaic (PV) Pumping Irrigation Systems With Storage." *Sustainability (Switzerland)* 12(3). https://doi.org/10.3390/su12031026.

Nguyen, Huu Nam, Quoc Thanh Tran, Canh Tung Ngo, Duc Dam Nguyen, and Van Quan Tran. 2025. "Solar Energy Prediction Through Machine Learning Models: A Comparative Analysis of Regressor Algorithms." *PLoS ONE* 20(1):1–23. https://doi.org/10.1371/journal.pone.0315955.

Quiles, Eduardo, Carlos Roldán-Blay, Guillermo Escrivá-Escrivá, and Carlos Roldán-Porta. 2020. "Accurate Sizing of Residential Stand-Alone Photovoltaic Systems Considering System Reliability." *Sustainability (Switzerland)* 12(3):1–18. https://doi.org/10.3390/su12031274.

Simon Haykin (McMaster University, Hamilton, Ontario, Canada). 2005. "Neural Networks—A Comprehensive Foundation—Simon Haykin.Pdf," 823.

Subramanian, E., M. Mithun Karthik, G. Prem Krishna, D. Vaisnav Prasath, and V. Sukesh Kumar. 2023. "Solar power prediction using machine learning." arXiv preprint arXiv:2303.07875.

Uhlir, Paul F., and Peter Schröder. 2007. "Open Data for Global Science." *Data Science Journal* 6(June):OD36–OD53. https://doi.org/10.2481/dsj.6.od36.

Zhang, Wenying, Huaguang Zhang, Jinhai Liu, Kai Li, Dongsheng Yang, and Hui Tian. 2017. "Weather Prediction With Multiclass Support Vector Machines in the Fault Detection of Photovoltaic System." *IEEE/CAA Journal of Automatica Sinica* 4(3):520–525. https://doi.org/10.1109/JAS.2017.7510562.

Chapter 6

Optimization of Solar Energy Using a Dual-Axis Solar Panel Tracking and Monitoring System

Saquibahmed Bagali and Venkatratnam Chitturi

6.1 INTRODUCTION

The demand for renewable energy is growing rapidly, with solar power being one of the most promising sources of clean energy. However, maximizing the efficiency of solar panels remains a significant challenge, as solar panels need to be positioned correctly to capture the maximum amount of sunlight throughout the day. Traditional solar panels, which are fixed in a single position, do not always receive optimal sunlight, especially during certain times of the day or under varying weather conditions. To address this challenge, this work focuses on the development of a "dual-axis solar panel tracking and monitoring system." This system uses advanced sensors and automation to optimize the position of a solar panel, ensuring that it follows the sun's movement and thus maximizing the energy output.

6.2 LITERATURE REVIEW

The transition to renewable energy is a cornerstone of global sustainable development, and solar energy has emerged as one of the most promising solutions to meet rising energy demands (Rekioua et al., 2023; Rubio et al., 2021). This literature review explores a collection of research on solar energy technologies, focusing on their optimization, challenges, and integration into sustainable systems (Sathish et al., 2023; Singh et al., 2024b).

6.2.1 Solar Energy Technologies

Integrating solar energy into global energy portfolios is not only a practical choice but also a pivotal step toward achieving long-term sustainability goals (Dutt et al., 2024). Solar energy is not just a source of electricity but a catalyst for sustainable development, offering multifaceted benefits across economic, social, environmental, and technological dimensions. Its continued expansion and integration are essential for achieving inclusive

DOI: 10.1201/9781003634737-6

and resilient development pathways worldwide (Ali, 2022). Optimization of these systems involves strategic considerations ranging from site selection and technology choice to operational efficiency and maintenance (Tiwari et al., 2022; Tomar et al., 2024). Location plays a critical role, with factors such as solar irradiance, shading, and geographic orientation influencing energy capture potential. Advancements in PV technologies, including improvements in cell efficiency and module design, contribute to maximizing electricity generation while minimizing costs (Pimoli et al., 2025). Integration of efficient inverters equipped with maximum power point tracking (MPPT) enhances overall system performance by optimizing the conversion of DC into AC suitable for grid integration. Additionally, smart system design, which minimizes shading and optimizes layout, further enhances output reliability. However, several challenges hinder the widespread adoption of solar PV. These include intermittency due to weather variations, which necessitate robust energy storage solutions and smart grid technologies to ensure a continuous power supply. Cost remains a barrier despite declining PV prices, requiring innovative financing mechanisms and supportive policies to achieve economic viability. Technological limitations, such as efficiency constraints and material availability, also pose challenges for further enhancing PV performance and reducing environmental impact. Overcoming these challenges demands collaborative efforts across industry, government, and research sectors to advance solar PV technology and realize its full potential as a sustainable energy solution. Ongoing research and development focus on improving efficiency, reducing costs, and addressing the challenges associated with solar PV optimization and integration. Solar photovoltaic (PV) optimization (Figure 6.1) and challenges are crucial aspects of harnessing renewable energy from the sun. Optimization techniques such as MPPT, panel cleaning and maintenance, system design and layout optimization, and inverter optimization aim to maximize energy production and reduce losses (Muhammad, 2022).

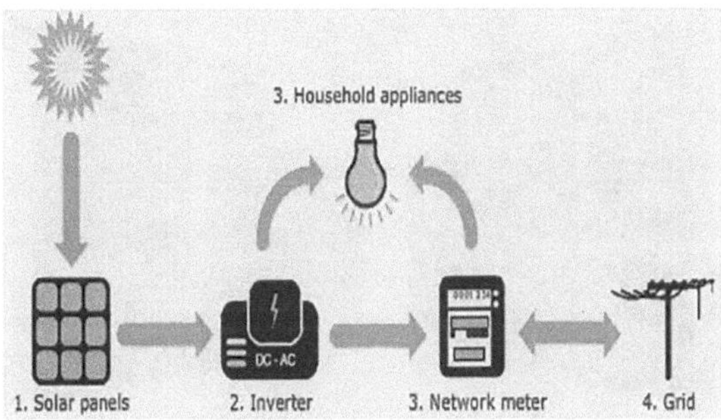

Figure 6.1 Solar photovoltaic energy optimization (Muhammad, 2022).

6.2.2 Solar Panel Types

Solar cells, also known as photovoltaic (PV) cells, are devices that convert light energy directly into electrical energy through the photovoltaic effect. The most commonly used material for solar cells is silicon, due to its favorable electronic properties and abundant availability. A typical silicon solar cell consists of a p-n junction, where a p-type semiconductor (positive) and an n-type semiconductor (negative) are brought together. When sunlight strikes the solar cell, photons with sufficient energy dislodge electrons from their atomic bonds in the semiconductor material, creating electron-hole pairs. These free electrons are then driven toward the n-type layer, while the holes move toward the p-type layer, creating an electric current. Solar cells are assembled into larger units called solar panels or modules, which are then integrated into photovoltaic systems. These systems can be deployed in a variety of scales and applications, from small rooftop installations to large solar farms (Khaled, 2022).

Monocrystalline solar panels, as shown in Figure 6.2, are one of the most efficient types of photovoltaic (PV) systems, known for their high efficiency and long lifespan. These panels are constructed using monocrystalline silicon, a highly pure form of silicon, which is derived from a single continuous crystal structure. The individual monocrystalline cells are then connected in series and parallel configurations to form a solar panel, also known as a module. This configuration enhances the voltage and current output, making the panels suitable for a wide range of applications, from residential rooftops to large-scale solar farms (Ansari & Al-Ezzi, 2022).

Polycrystalline solar panels, as shown in Figure 6.3, also known as multicrystalline panels, are a type of photovoltaic technology used to convert sunlight into electricity. These panels are constructed from silicon crystals, which are melted together and then cooled to form a solid block or ingot. These are known for their cost-effectiveness and ease of production compared to their monocrystalline counterparts. Although they are generally

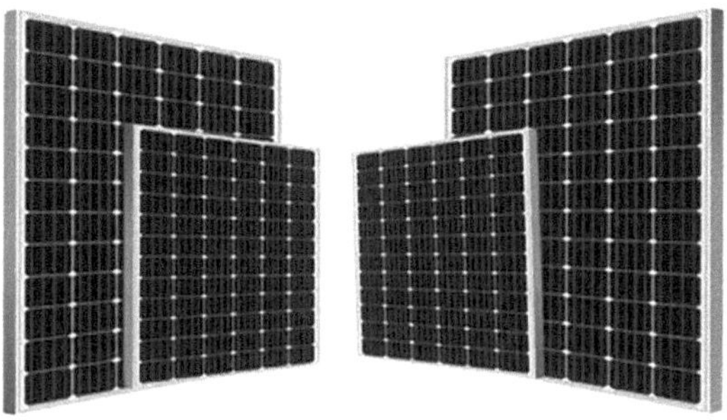

Figure 6.2 Monocrystalline solar panels (Ansari & Al-Ezzi, 2022).

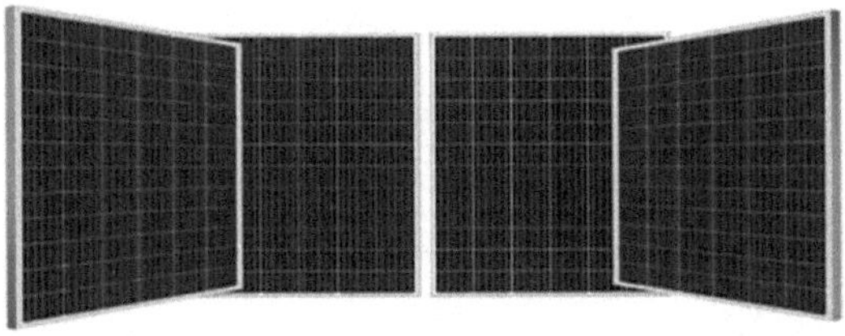

Figure 6.3 Polycrystalline solar panels (Ansari & Al-Ezzi, 2022).

Figure 6.4 Thin film solar panels (Ansari & Al-Ezzi, 2022).

less efficient, with typical efficiencies ranging from 15% to 17%, their lower manufacturing costs make them an attractive option for residential and commercial solar installations. Additionally, their performance in various environmental conditions and their robustness contribute to their widespread use in the solar energy market (Ansari & Al-Ezzi, 2022).

Thin-film solar panels shown in Figure 6.4 represent a diverse category of photovoltaic technologies characterized by their lightweight and flexible construction. Unlike traditional crystalline silicon panels, thin-film panels are made by depositing one or more layers of photovoltaic material onto a substrate, such as glass, plastic, or metal. Thin-film solar panels offer several advantages, including flexibility, lightweight construction, and the ability to perform well in low-light conditions and high temperatures. They can be integrated into a variety of surfaces, including building facades, windows, and even flexible applications like solar backpacks or tents. Despite their generally lower efficiency compared to crystalline silicon panels, which typically range from 10% to 12%, their unique properties and lower production costs make them an attractive option for specific applications where flexibility and weight are crucial considerations (Ansari & Al-Ezzi, 2022).

Concentrated solar panels, or concentrated photovoltaic (CPV) systems, shown in Figure 6.5, are designed to enhance the efficiency of photovoltaic cells by focusing sunlight onto a smaller, high-efficiency solar cell area using optical devices such as lenses or mirrors. CPV systems are particularly

Figure 6.5 Concentrated solar panels (Ansari & Al-Ezzi, 2022).

suited for areas with high direct normal irradiance (DNI), such as deserts and other sunny regions, where they can generate large amounts of electricity efficiently. However, their reliance on direct sunlight means they are less effective in cloudy or diffuse light conditions compared to traditional solar panels. Despite this limitation, the high efficiency and potential for reduced material costs make CPV systems an attractive option for large-scale solar power installations aiming to maximize energy output from available land (Ansari & Al-Ezzi, 2022).

Solar panel efficiency refers to the percentage of sunlight that a solar panel can convert into electricity under standard test conditions (Rubino et al., 2021). Solar cell efficiency (maximum):

$$\eta_{max} = \frac{P_{max}}{E \times A_c} \qquad \ldots(6.1)$$

where, P_{max} = maximum power output (in W); E = incident radiation flux (in W/m^2) and A_c = area of collector (in m^2)

6.2.3 Sensors Used in Solar Power Systems

Some of the common sensors used in solar power systems include the following (Dharmateja & Bevi, 2021):

- ACS712 Current Sensor: This sensor uses the Hall effect to measure both AC and DC currents. It provides an analogue voltage output proportional to the current passing through it. The ACS712 is widely used in motor control, battery monitoring, and power management applications due to its ease of use and precision.

- 0–25V Voltage Sensor Module: This sensor is designed to measure voltages in the range of 0 to 25 volts. It typically uses a voltage divider circuit to scale down the input voltage to a level that can be read by a microcontroller's analogue-to-digital converter (ADC). It is often used in battery monitoring, power supply measurement, and other applications where voltage measurement is required.
- Resistance Temperature Detectors (RTDs): These use materials, typically platinum, whose resistance changes with temperature. RTDs are known for their accuracy and stability.
- Piezoelectric Sensors: These generate an electric charge in response to mechanical stress. They are widely used in dynamic pressure measurements.
- LDR (Light-Dependent Resistors): Change resistance based on light intensity and is common in street lighting and night light.

Real-time monitoring of solar power plants has become an essential component for maximizing the performance, reliability, and efficiency of solar energy systems. This advanced monitoring leverages sophisticated sensors and IoT (Internet of Things) technology to continuously track critical parameters such as voltage, current, power output, and overall system efficiency. By transmitting this data to a central server, the information can be accessed in real-time via smartphones, web portals, and email notifications, providing comprehensive oversight and immediate feedback to stakeholders. Furthermore, historical data analysis helps in understanding the long-term performance of the solar power plant. By examining past performance trends, operators can gain insights into seasonal variations, degradation rates, and the impact of different maintenance activities. This information is invaluable for making informed decisions about future investments, upgrades, and maintenance strategies. It also supports compliance with performance guarantees and contracts, ensuring that the solar power plant meets its expected output over its operational lifespan (Aung & Zaw, 2023).

One of the significant benefits of an IoT-based solar tracker with power monitoring is remote accessibility and control. The collected data is accessible via web interfaces, smartphone applications, and email notifications, enabling operators and stakeholders to monitor the system's performance from anywhere at any time. This remote monitoring capability allows for proactive maintenance and quick response to any issues, minimizing downtime and ensuring continuous optimal performance. Moreover, the system can generate automated reports and alerts. For example, if the power output falls below a certain threshold, the system can send an immediate alert to the maintenance team, prompting them to investigate and resolve the issue. Historical data analysis further aids in predictive maintenance, allowing operators to anticipate and address potential problems before they lead to significant performance drops.

The IoT-based solar tracker also contributes to better energy management and planning. By providing real-time and historical data on energy production, the system helps in forecasting energy yields and managing energy storage and distribution more effectively. This is particularly valuable for grid-tied solar systems and hybrid setups, where efficient energy management is crucial for balancing supply and demand. The environmental impact of such a system is also noteworthy. By maximizing the efficiency of solar energy capture, the IoT-based solar tracker helps in reducing the reliance on fossil fuels and lowering greenhouse gas emissions. This aligns with global efforts to promote sustainable energy solutions and combat climate change (Maulana & Ardhenta, 2022).

DIY solar panels monitoring system, Figure 6.6, can be quite affordable and manageable, even for those with limited technical expertise. Basic components like sensors, microcontrollers, and display modules are readily available and inexpensive. Numerous online resources, tutorials, and community forums provide step-by-step guides and support for setting up and troubleshooting your system. Building a DIY monitoring system also offers educational benefits, giving you a deeper understanding of how solar panels work and the factors that influence their performance. It can be a great learning project for hobbyists, students, or anyone interested in renewable energy technologies. Moreover, the skills gained from this project can be applied to other DIY electronics and IoT projects, broadening technical knowledge and capabilities (Ramkiran et al., 2020).

Extensive research has been carried out using different color filter papers to evaluate the electrical performance of the solar photovoltaic module. Five different filters were used from magenta to red so that the relationship of frequency or wavelength and photonic energy can be determined. The

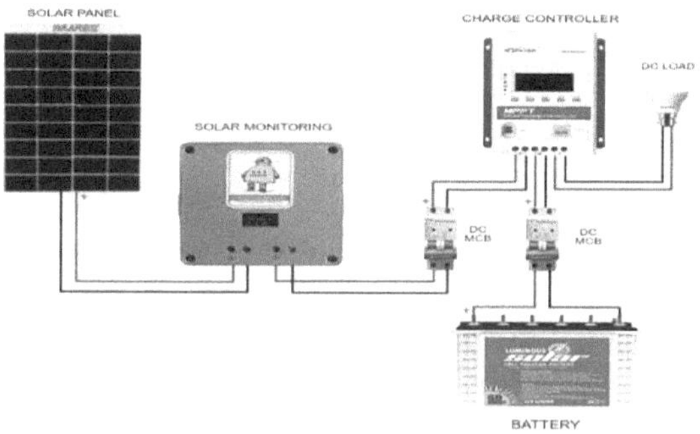

Figure 6.6 DIY solar panel monitoring system (Ramkiran et al., 2020).

efficiency and power of the solar photovoltaic module were measured with and without filters, and their performance was compared. From the study, it has been concluded that among all the color filters used, magenta color in the visible spectrum of solar radiation gave the maximum efficiency. Results also show that blue and magenta have maximum scope for producing more change in temperature when compared to other filters.

The equipment needed for the state-of-the-art setup is described in this section. Two 50 Wp power capacity panels were used for performing this experiment. It was oriented at optimum tilt equal to the latitude of the place for maximum exposure to sunlight. A 3W lamp load was connected to the PV Module for measuring the VI characteristics. Low-cost cellophane color filters are used to cover the photovoltaic cell in the real outdoor condition of the full solar spectrum. The filters of different colors (red, yellow, green, blue, and magenta) were used in the experiment. These filter papers are opaque to the frequency of light corresponding to their own color. Furthermore, the thermo-physical properties, optical properties, and thicknesses of the filters were not known in the aforementioned work. All the color filters were placed, and the measurements were taken and compared. The output of two modules with and without filters is measured individually (Kannan & Vakeesan, 2022; Bhandari et al., 2023).

6.2.4 Solar Panel Tracking Systems

Solar tracking systems have gained significant attention as a means to improve the efficiency of solar power generation. These systems adjust the orientation of solar panels or photovoltaic (PV) arrays to follow the movement of the Sun, increasing the amount of solar radiation absorbed and thus enhancing the energy yield. Solar trackers are generally classified into single-axis and dual-axis types. Single-axis trackers, which rotate along one axis (either horizontally or vertically), are simpler and cost-effective, with horizontal single-axis trackers being more common for locations near the equator. Dual-axis trackers, offering two degrees of freedom, allow panels to follow the Sun's movement both horizontally and vertically, providing higher energy yield but at the cost of increased complexity and maintenance. Hybrid systems, combining the features of both, have also been developed to optimize performance while minimizing drawbacks. Research in this area has led to various advancements, such as novel control strategies for single-axis trackers that reduce mechanical friction and improve response time, which can boost energy output by up to 25%. Dual-axis trackers, despite their complexity, can increase energy production by as much as 40% compared to fixed systems, especially when paired with advanced tracking algorithms. However, these benefits come with increased costs, in terms of both installation and maintenance. The higher initial investment is often offset by long-term energy production gains, particularly in regions with high solar insolation. Studies have also explored cost-effective solutions, such as low-cost materials and simplified designs, which help make solar tracking systems more economically viable. Environmental impacts are generally positive, as solar trackers increase energy capture, thereby reducing the overall carbon footprint of solar installations. Nonetheless, the mechanical components of tracking systems contribute to environmental costs, particularly during production and maintenance. The main challenges include mechanical reliability, as the motors and sensors of tracking systems are subject to wear and tear, and the adaptation of the systems to various geographic and climatic conditions. High winds, snow, and dust accumulation can hinder performance, necessitating features like automatic cleaning or enhanced wind resistance. Future research is focusing on improving the durability of trackers, incorporating predictive maintenance, and enhancing the integration of tracking systems with smart grids for better energy distribution. In conclusion, while solar tracking systems offer substantial efficiency improvements, they are accompanied by higher costs and maintenance requirements. Ongoing innovations in cost reduction, system reliability, and integration with energy management technologies are crucial for making solar tracking systems a more sustainable and widely adopted solution in the renewable energy sector (Beltran et al., 2020).

Solar tracking systems improve the efficiency of solar energy generation by adjusting the orientation of solar panels to follow the Sun's movement. These systems can be classified into different types based on the drive mechanisms (active or passive) and the number of axes (single-axis or dual-axis). Each type has its advantages and disadvantages, which affect their performance, cost, and maintenance.

Active trackers use motors, actuators, and sensors to move solar panels, adjusting their orientation to optimize energy capture throughout the day. These trackers offer high accuracy and efficiency, with energy output increasing by up to 40% compared to fixed systems. Active trackers are ideal for large-scale commercial applications and locations with significant solar variability. However, they come with higher costs and maintenance due to the reliance on mechanical components that may wear out over time. According to Martín-Utrilla et al. (2022), active systems are more effective but require frequent maintenance to ensure optimal performance, especially in harsh environments.

Passive trackers, Figure 6.7, do not require external power sources. They rely on natural forces like gravity, heat, or hydraulic pressure to move the panels. These systems are simpler, less expensive, and require less maintenance. While they are less efficient than active trackers, passive systems can still improve energy production by 15–25% compared to fixed panels. However, passive trackers are slower and less accurate, making them less effective in locations with rapidly changing weather conditions.

Figure 6.7 Passive trackers (Panait, 2022).

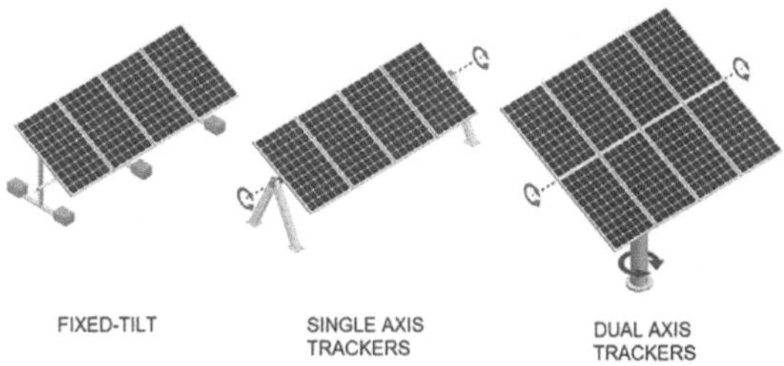

Figure 6.8 Single and dual-axis trackers (Panait, 2022).

Single-axis trackers, Figure 6.8, rotate the solar panels along one axis, typically following the Sun's east-to-west movement. These systems are commonly used due to their simplicity, lower cost, and moderate energy yield improvement (20–30%) over fixed systems. It was noted that horizontal single-axis trackers are ideal for equatorial regions, while vertical trackers work better in high-latitude areas. They provide an efficient balance between cost and energy output, though their performance drops during winter when the Sun's angle changes drastically.

Dual-axis trackers, Figure 6.8, move the panels along both horizontal and vertical axes, allowing them to follow the Sun's path more precisely. These systems offer the highest efficiency, increasing energy capture by up to 50% compared to fixed panels. However, dual-axis systems are more complex and expensive to install and maintain due to the additional moving parts and sensors. They are most suitable for regions with significant seasonal variations in solar radiation, where maximum energy capture is essential.

Solar trackers, whether active or passive, single-axis or dual-axis, offer distinct benefits depending on location and application. Active trackers provide higher efficiency but require more maintenance and higher initial investment. Passive trackers, while cheaper and simpler, are less precise and offer lower energy gains. Single-axis trackers strike a balance between cost and performance, making them suitable for many applications, while dual-axis trackers provide the highest efficiency but at a higher cost and complexity. Ongoing advancements in materials, design, and control algorithms are focused on improving the affordability and reliability of solar tracking systems, making them an increasingly viable option for enhancing solar energy production.

In summary, solar tracking systems can be classified based on drive types and axis types, each offering distinct advantages and limitations. Active trackers provide high efficiency and precision but at the cost of higher complexity and maintenance. Passive trackers, while simpler and more cost-effective, typically offer lower energy gains and reduced accuracy. In terms of axis types, single-axis trackers are widely used for their cost-effectiveness and moderate energy gains, while dual-axis trackers are more suitable for high-latitude regions or areas with significant solar variability, offering the highest potential energy yield but at a higher cost and complexity. The choice between these systems depends on factors such as location, budget, and energy goals, with ongoing advancements focusing on improving the cost-effectiveness, reliability, and integration of these systems into modern solar power infrastructures (Singh et al., 2024a).

6.2.5 Efficiency of Solar Tracking Systems

The primary benefit of solar tracking systems is the increased energy output compared to fixed solar panels. This gain is attributed to the ability of tracking systems to maintain an optimal angle of incidence, which maximizes solar radiation absorption. Solar tracking systems can increase energy output by approximately 20–40% for single-axis trackers and up to 50% for dual-axis trackers, depending on the geographical location and the type of tracker.

The efficiency of a solar tracking system can be expressed using the tracking gain or efficiency gain (η), which represents the increase in energy production compared to a fixed solar panel system. The formula for calculating tracking gain is:

$$\text{Effiency} = \frac{x - y}{x} * 100 \qquad \ldots.(6.2)$$

where, X = Power with tracker and Y = Power without tracker

Tracking gain is often used to compare the energy produced by different tracking systems, and its value can vary based on factors such as panel orientation, location, and weather conditions (Hamilton, 2021).

6.3 METHODOLOGY

The block diagram of the dual-axis solar panel tracking and monitoring system shown in Figure 6.9 provides a clear representation of the key components and their interconnections. At the core of the system is the Arduino Uno microcontroller, which acts as the central processing unit. It processes input signals from various sensors and sends control commands to actuators. The system integrates LDRs, strategically positioned to detect the intensity of sunlight from different directions. These LDRs provide analogue signals to the Arduino, enabling it to determine the optimal position of the solar panel.

The servo motors are connected to the Arduino and are responsible for adjusting the panel's orientation in both horizontal and vertical axes based on the LDR inputs. This dual-axis mechanism ensures that the panel continuously aligns with the sun, maximizing energy capture. The solar panel, mounted on a movable frame controlled by the servos, converts sunlight into electrical energy, which is monitored by a set of sensors. The monitoring system includes an ACS712 current sensor and a voltage sensor, which measure the electrical output of the solar panel in real-time. Additionally, a DHT22 sensor is integrated to monitor environmental parameters such as temperature and humidity. These sensors feed data into the Arduino, which processes and outputs the

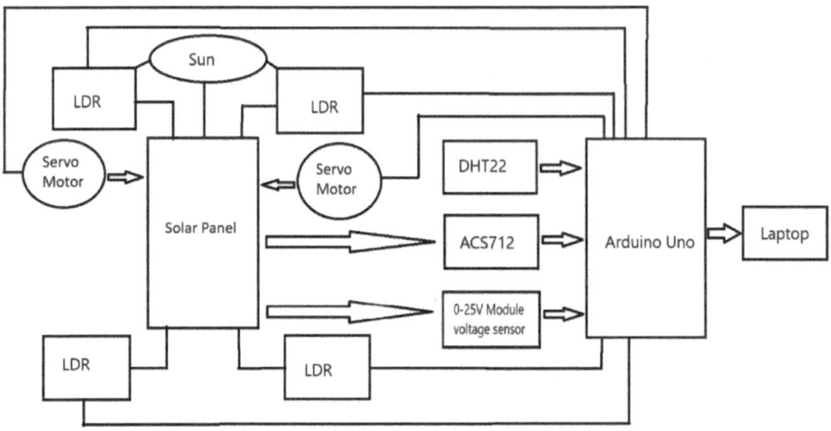

Figure 6.9 Block diagram of the proposed system.

Figure 6.10 Prototype of the solar power monitoring system.

information to either a serial monitor or an optional LCD display for real-time monitoring. A power supply unit ensures the system components receive adequate power for operation. The entire setup is shown in Figure 6.10.

6.4 RESULTS AND DISCUSSION

This section evaluates the performance of the proposed design, identifies discrepancies between theoretical and experimental results, and explores potential sources of error. Additionally, troubleshooting methods to address these issues are outlined to ensure accurate and reliable operation.

6.4.1 Low Light Conditions

Figure 6.11 shows the operation of the solar panel in low light conditions, and the corresponding readings along with the graph are shown in Figure 6.12.

6.4.2 Moderate Light Conditions

Figure 6.13 shows the operation of the solar panel in low light conditions, and the corresponding readings along with the graph are shown in Figure 6.14.

Figure 6.11 Operation in low light conditions.

Sl.No	Fixed Panel (W)	Dual Axis (W)
1	0	0
2	0	0
3	0	0
4	0.1	0.12
5	0.15	0.18
6	0.25	0.3
7	0.35	0.42
8	0.4	0.5
9	0.5	0.6
10	0.6	0.73

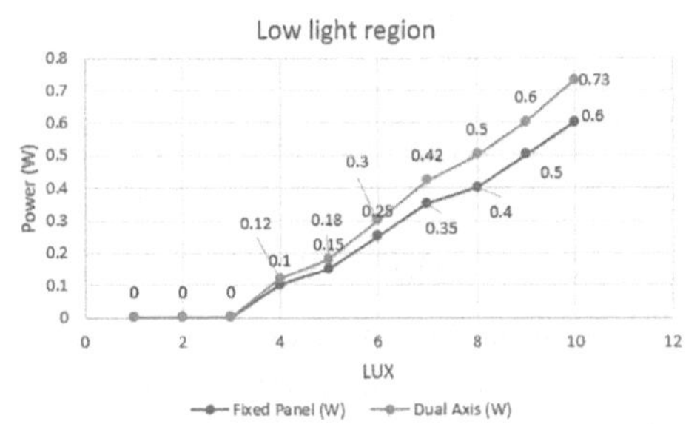

Figure 6.12 Sample data and graph for low light conditions.

Figure 6.13 Operation in moderate light conditions.

Sl.No	Fixed Panel (W)	Dual Axis (W)
1	1.91	2.5
2	1.82	2.41
3	1.93	2.51
4	1.95	2.55
5	2	2.6
6	2.05	2.71
7	2.12	2.8
8	2.15	2.87
9	2.23	2.9
10	2.25	2.94

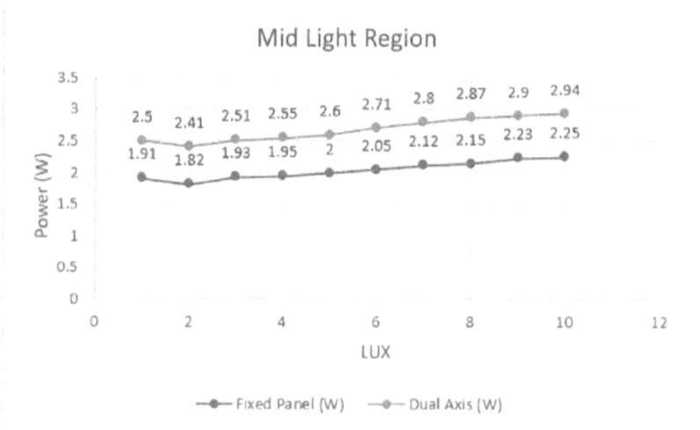

Figure 6.14 Sample data and graph for moderate light conditions.

6.4.3 Good Light Conditions

Figure 6.15 shows the operation of the solar panel in low light conditions, and the corresponding readings along with the graph are shown in Figure 6.16.

The observed discrepancies between theoretical and experimental results are attributed to various factors, like sensor inaccuracy due to electrical

Figure 6.15 Operation in good light conditions.

Sl.No	Fixed Panel (W)	Dual Axis (W)
1	2.86	3.85
2	2.92	3.94
3	2.79	3.81
4	2.95	4
5	3	4.05
6	3.03	4.14
7	3.1	4.22
8	3.18	4.3
9	3.21	4.34
10	3.28	4.39

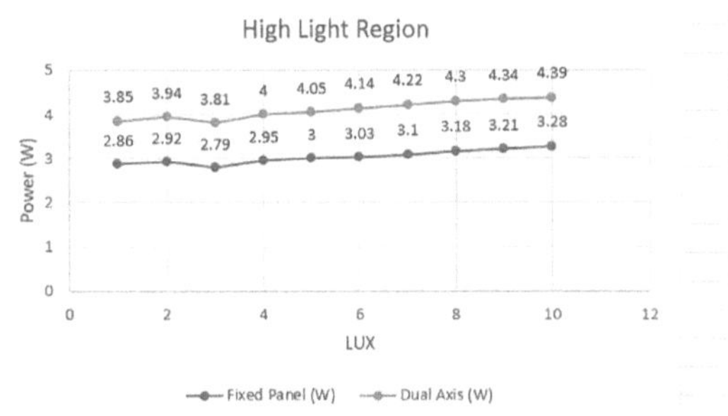

Figure 6.16 Sample data and graph for good light conditions.

noise, that is, noise in the circuit interfered with sensor readings (noise-filtering components such as capacitors and ferrite beads to stabilize sensor outputs); calibration errors: sensors like ACS712 and voltage sensors were not perfectly calibrated, causing deviations (regularly calibration against standard measuring devices to ensure accuracy); servo motor response time was delayed as Arduino took extra time to compare sensor data and send control signals and mechanical limitations of servo motors exhibiting backlash or inertia, slowing their response (replaced the servo motors with higher-torque, faster-response models for precise movement).

6.5 CONCLUSION AND FUTURE SCOPE

The core idea in this chapter was to use dual-axis tracking, which enables the solar panel to rotate both horizontally (azimuth) and vertically (elevation). This allows the panel to always face the sun, unlike fixed-position panels, which may be inefficient during certain times of the day. By using dual-axis tracking, the system is capable of adjusting the panel's orientation throughout the day, ensuring that it remains in the optimal position to capture the most sunlight. To achieve this, the system integrates a range of sensors and components.

The DHT22 sensor, which measures temperature and humidity, plays a crucial role in tracking environmental factors that can influence the efficiency of the solar panel. Temperature, in particular, affects the performance of solar panels, and by monitoring it, the system can help prevent overheating, which can degrade the efficiency of the solar cells. The ACS712 current sensor is used to monitor the current generated by the solar panel, providing real-time feedback on the amount of energy being produced. This is crucial for assessing the efficiency of the system and ensuring that the solar panel is functioning optimally. In addition to current monitoring, the system incorporates a 0–25V voltage sensor, which tracks the voltage output from the solar panel. This allows for monitoring the electrical performance and ensures that the system operates within the correct voltage range for optimal energy harvesting. Light Dependent Resistors (LDRs) are utilized to measure the intensity of sunlight. The readings from the LDRs allow the system to determine the position of the sun, and this data is used to calculate the most efficient angle for the solar panel orientation via MG995 servo motors.

The combination of dual-axis tracking and real-time monitoring makes this system highly efficient and reliable. This not only improves energy efficiency but also allows for constant monitoring of the solar panel's performance. Moreover, the system provides a scalable solution for renewable energy applications. By implementing this dual-axis tracking and monitoring system in larger solar installations, energy production can be significantly enhanced, leading to better utilization of solar power. In the future, similar systems could be integrated with smart grids or energy storage solutions, creating more advanced and sustainable energy systems for the future.

REFERENCES

Ansari, M. N. M., & Al-Ezzi, A. S. (2022). Photovoltaic solar cells: A review. *Applied System Innovation*, 5(4), 67. https://doi.org/10.3390/asi5040067

Aung, Y. W., & Zaw, N. W. (2023). Implementation of solar photovoltaic data monitoring system. *International Journal of Science*, 7(8).

Beltran, J. A., Gonzalez Rubio, J. L. S., & Garcia-Beltran, C. D. (2020). *Design, manufacturing and performance test of a solar tracker made by an embedded control*. CERMA.

Bhandari, P., Padalia, D., Ranakoti, L., Khargotra, R., András, K., & Singh, T. (2023). Thermo-hydraulic investigation of open micro prism pin fin heat sink having varying prism sides. *Alexandria Engineering Journal*, 69, 457–468. https://doi.org/10.1016/j.aej.2023.02.016.

Dharmateja, R., & Bevi, A. R. (2021). The review mainly focuses on different types of sensors and its technology. *International Journal on Recent and Innovation Trends in Computing and Communication*, 3(2).

Dutt, N., Hedau, A., Kumar, A., Awasthi, M. K., & Singh, V. P. (2024). Effect of duct length variation on solar air heater performance for smooth and D-shaped roughened absorber plate. *Heat Transfer*. https://doi.org/10.1002/htj.23115.

Hamilton, S. J. (2021). Sun-tracking solar cell array system. Thesis, University of Queensland Department of Computer Science and Electrical Engineering.

Kannan, N., & Vakeesan, D. (2022). Solar energy for future world: A review. *Renewable and Sustainable Energy Reviews*, 62, 1082–1125.

Khaled, A. (2022). Research and modeling of hybrid energy systems (review). *Eco Energetics*, 1, 65–69.

Martín-Utrilla, F. D., Chaves-Ávila, J. P., & Cossent, R. (2022). Decision framework for selecting flexibility mechanisms in distribution grids. *Economics of Energy & Environmental Policy*, 11(2).

Maulana, R. K. S., & Ardhenta, L. (2022). Sensors and their applications: Simplifying technology for everyday use. *Indonesian Journal of Electrical Engineering and Computer Science*, 10.

Muhammad, A., et al. (2022). Solar photovoltaic energy optimization and challenges. *Frontiers in Energy Research*, 10.

Panait, T. T. (2022). *A simple neural network solar tracker for optimizing conversion efficiency in off-grid solar generators*. ICREPQ.

Pimoli, D. S., Tomar, P., Kumar, A., Dwivedi, G., & Shukla, A. K. (2025). Experimental investigation of friction factor and heat transfer enhancement using circular perforated twisted tape inserts in heat exchangers. *International Journal of Energy for a Clean Environment*. https://doi.org/10.1615/InterJEnerCleanEnv.2024053616.

Ramkiran, B., Sundarabalan, C. K., & Sudhakar, K. (2020). Performance evaluation of solar PV module with filters in an outdoor environment. *Case Studies in Thermal Engineering*, 21, 100700.

Rekioua, D., Mokrani, Z., Kakouche, K., Rekioua, T., Oubelaid, A., Logerais, P. O., . . . Ghoneim, S. S. (2023). Optimization and intelligent power management control for an autonomous hybrid wind turbine photovoltaic diesel generator with batteries. *Scientific Reports*, 13(1), 21830. http://dx.doi.org/10.1038/s41598-023-49067-4.

Rubino, F., Poza, P., Pasquino, G., & Carlone, P. (2021). Thermal spray processes in concentrating solar power technology. *Metals*, 11(9), 1377.

Rubio, F. R., Ortega, M. G., Gordillo, F., & López-Martínez, M. (2021). Application of new control strategy for sun tracking. *Energy Convers Manage*, 48(7), 2174–2184.

Sathish, T., Saravanan, R., Jeyanthi, C. E., Sabariraj, R. V., Hemachandu, P., Ağbulut, Ü., . . . Soudagar, M. E. M. (2023). PCM embedded square channel receiver for effective operation of net zero emission/energy PDC building heating system. *Sustainable Energy Technologies and Assessments*, 60, 103519. http://dx.doi.org/10.1016/j.seta.2023.103519.

Singh, A. R., Kumar, R. S., Bajaj, M., Khadse, C. B., & Zaitsev, I. (2024a). Machine learning-based energy management and power forecasting in grid-connected microgrids with multiple distributed energy sources. *Scientific Reports*, 14(1), 19207. https://doi.org/10.1038/s41598-024-70336-3.

Singh, B., Kaur, P., Yadav, A. K., Awasthi, M. K., & Kumar, A. (2024b). Design and analysis of solar tracking system for PV thermal performance enhancement. In *Heat Transfer Enhancement Techniques: Thermal Performance, Optimization and Applications*, A. Kumar, N. Dutt, & M. K. Awasthi, Eds. John Wiley & Sons, 2024, pp. 251–267, https://doi.org/10.1002/9781394270996.ch11.

Tiwari, S., Swaminathan, M., Harender, Eashwar, S., & Singh, D. B. (2022). Performance enhancement of the photovoltaic system with different cooling methods. *Environmental Science and Pollution Research*, 29(30), 45107–45130. https://link.springer.com/article/10.1007/s11356-022-20330-x.

Tomar, P., Chauhan, P., Kumar, A., & Shukla, A. K. (2024). Numerical investigation of thermal performance enhancement of solar reservoir using flash cycle. *Journal of Advanced Research in Fluid Mechanics and Thermal Sciences*, 123(1), 197–221. https://doi.org/10.37934/arfmts.123.1.197221.

Chapter 7

Optimized Design and Modeling of a 4kW Hybrid Solar EV Charging Station

Abhilasha Pawar, Amit Kumar Roy, Ashish Tewari, Divyansh Tiwari, Ishita Singh, and Rishu Pandey

7.1 INTRODUCTION

The rising demand for fuel, particularly automobiles, has led to an increase in CO_2 emissions, which contribute to environmental problems. One of the most pressing challenges today is finding ways to reduce these emissions, and electric vehicles (EVs) offer a promising solution. As the world moves toward green mobility, people are motivated to transition to electric vehicles (EVs) to help reduce pollution. However, the rapid adoption of EVs comes with its own set of challenges—especially when it comes to charging. Rapid and ultra-fast charging stations can exert considerable strain on the power grid, leading to issues like power surges, voltage drops, and supply shortages. One way to overcome these challenges is by integrating renewable energy sources and battery energy storage systems (BESS) into hybrid charging stations. These stations utilize solar energy and battery storage to ensure continuous EV charging while alleviating stress on the grid. Solar photovoltaic (PV) panels are a clean and increasingly affordable way to generate electricity directly from sunlight, without harming the environment.

In recent years, EV use has grown, and finding convenient ways to charge them has become a priority for drivers. Simultaneously, the declining costs of solar PV systems have made them a viable option for power generation. This project aims to develop a charging station that integrates solar energy and battery storage to supply power to EVs. When solar power isn't available—such as during cloudy weather or at night—the station can also draw power from the electrical grid. The system would be smart enough to manage energy flows between the solar panels, battery storage, grid, and vehicles. The solar panels produce sufficient electricity. During daylight hours to charge the EVs, and any extra energy can be sent back to the grid. When sunlight is limited, the station will rely on stored battery power or tap into the grid if necessary.

To make sure the system operates efficiently, a Perturb and Observe (P&O) algorithm will be implemented for optimization of the power output of the solar panels. A converter will also be included to manage energy flow

DOI: 10.1201/9781003634737-7

between the system and a single-phase AC grid, ensuring that everything works smoothly. This project will be modeled in MATLAB® to simulate real-world conditions, ensuring that the charging station can provide continuous power for EVs while minimizing its impact on the grid. It offers a practical solution to some of the key challenges in creating a sustainable and reliable EV charging infrastructure.

7.2 LITERATURE REVIEW

This text elaborates on the complexities of photovoltaic (PV) power systems, focusing on how to optimize power tracking and fast-charging stations for electric vehicles (EVs) (Tan *et al.*, 2020; Chandra *et al.*, 2016).

Due to the weather-dependent nature of solar energy, a renewable energy-powered fast-charging station for EVs must be supported by battery storage (Badri *et al.*, 2023; Shariff *et al.*, 2020). This setup typically pairs a grid-interactive PV system with a battery to ensure a continuous energy supply, essential for high-speed charging. To achieve affordability, reliability, and effectiveness, improved designs for these stations are constantly being developed.

The system utilizes the Perturb and Observe (P&O) algorithm, which ensures the photovoltaic (PV) panels function at their peak power point (MPP) (Singla *et al.*, 2024). The battery is charged in three phases: MPPT buck charging, constant voltage absorption stage, and float charging. The P&O algorithm is widely favored for its simplicity, requiring minimal sensors and circuitry. However, challenges like partial shading, which cause mismatched power output from different parts of the PV array, can limit performance. Particle Swarm Optimization (PSO) is another approach that has been proposed to enhance MPP tracking, particularly under such conditions (Sivasami & Khan, 2023). PSO is more efficient than P&O in certain cases, offering better adaptability.

Other methods such as Fuzzy Logic Controllers (FLC) or Artificial Neural Networks (ANN) provide alternative MPPT strategies, offering robustness under varying conditions. FLC monitors the MPP using logic rules, while ANN can be trained to further optimize performance based on data from FLC. To transform the DC power generated by the PV system into AC for grid usage, an inverter is employed, often regulated by Pulse Width Modulation (PWM) techniques (Shafiq *et al.*, 2023). The inverter's output needs to be synchronized with the grid using Phase Locked Loop (PLL) systems for matching the frequency and phase. Harmonics are filtered out using an LC filter, and transformers can ensure proper voltage and isolation for grid integration.

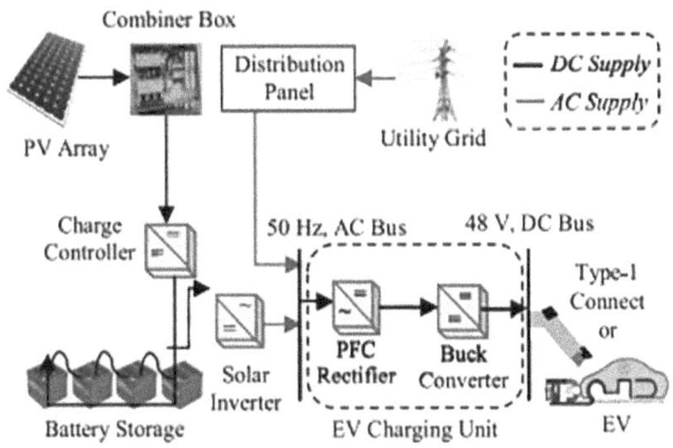

Figure 7.1 Block diagram.

The effectiveness of these systems depends on selecting the right MPPT method, as numerous techniques like Incremental Conductance, Sliding Mode Control (SMC), and Grey Wolf Optimization (GWO) have been explored. Under real-world conditions, like partial shading, conventional methods may struggle due to the presence of multiple peaks in the power curve, which more advanced techniques aim to overcome (Singh *et al.*, 2023, 2024a; Ramakrishnan *et al.*, 2024; Kumar *et al.*, 2024; Aggarwal *et al.*, 2024; Ravindran *et al.*, 2023).

Each of these strategies has its own trade-offs, and selection depends on factors like complexity, cost, and ability to handle various operational challenges (Singh *et al.*, 2024b; Bansal *et al.*, 2024; Upadhyay *et al.*, 2024; Singh and Kumar, 2024). The basic block diagram for the simulation design of the charging station has been shown in Figure 7.1.

7.3 METHODOLOGY

7.3.1 Solar-Powered EV Charging System with a DC-DC Converter

DC-DC converter plays an important role in transferring energy from solar panels to electric vehicle (EV) charging systems (Geetha & Usha, 2022). These converters adjust the input voltage to match the requirements of the EV battery. Typically, the voltage generated by photovoltaic (PV) cells is higher than what the battery requires, making it necessary to use a buck converter, also known as a step-down converter. Buck converters are widely employed in solar-powered charging systems due to their ability to efficiently lower the input voltage to the desired level.

7.4 OPERATION OF A BUCK CONVERTER

In solar-powered EV charging systems, the buck converter reduces the high voltage output from the solar array to a level suitable for charging the EV Battery. As it lowers the voltage, it simultaneously increases the current to ensure the battery receives adequate charging (Pinjari *et al.*, 2016).

Key Components of Buck Converter

A typical buck converter consists of the following components:

- **MOSFET Switch:** Controls the flow of energy through the system.
- **High-Power Inductor:** Maintains a steady flow of current.
- **Schottky Diode:** Ensures current flow when the switch is off and has a low forward voltage drop (typically around 0.5 V).
- **Capacitors:** Located at the input and output, these stabilize voltage and filter out any current ripple.
- **Reverse Blocking Diode:** Prevents current from flowing back to the PV panels, especially during low or no solar output (e.g., at night).

In operation, the MOSFET is designed to have minimal on-resistance (e.g., 0.02 Ω) and operates at a switching frequency of 1,000 Hz, controlled by a pulse-width modulation (PWM) signal. This high-frequency switching allows for precise regulation of the output voltage.

7.5 OUTPUT VOLTAGE CONTROL BY DUTY CYCLE

Output voltage of buck converter is given the duty cycle of the PWM signal, which refers to the proportion of each switching cycle during which the MOSFET is turned on. The relationship is given by Equation 7.1:

$$\text{Vout} = \text{D} \times \text{Vin} \tag{7.1}$$

Here, D represents the duty cycle, and Vin is the input voltage from solar array.

7.6 MANAGING RIPPLE IN CURRENT AND VOLTAGE

To ensure stable power delivery, the design of the buck converter accounts for ripple, which refers to small fluctuations in current and voltage. This is achieved by calculating the size of the key components:

- **Inductor Ripple Current:** Managed to maintain steady current flow, calculated as:

$$\Delta \mathrm{IL} = \frac{Vin \times D \times (1-D)}{fsw \times L} \tag{7.2}$$

where ΔIL is inductor ripple current, fsw is the switching frequency (1,000 Hz), D is duty cycle, and L is inductor value.

- **Output Capacitor Ripple Voltage***:* The capacitor smooths voltage fluctuations, calculated as:

$$\Delta \mathrm{VC} = \frac{Vin \times D \times (1-D)}{8 \times fsw \times C} \tag{7.3}$$

where ΔVC is capacitor ripple voltage and C is capacitance.

These design considerations ensure efficient energy transfer and stable charging performance in solar-powered EV systems.

7.7 MAXIMUM POWER POINT TRACKING (MPPT) WITH THE PERTURB AND OBSERVE (P&O) METHOD IN SOLAR CHARGING

"The **Maximum Power Point Tracking** (**MPPT**) algorithm is a vital component in solar electric vehicle (EV) charging systems" (Singh *et al.*, 2021). Its primary role is to ensure that solar panels work at the most efficient power level, adapting dynamically to changes in sunlight and temperature. Out of many MPPT techniques, the Perturb and Observe (P&O) method stands out as a popular choice, particularly for small- to medium-scale solar PV systems. This is due to its design, effectiveness in optimizing power capture, and ease of implementation, making it especially suitable for solar EV charging applications.

7.8 UNDERSTANDING THE P&O ALGORITHM

The P&O algorithm operates by making small adjustments to the PV array's voltage and observing how these changes affect power output. This process involves slightly altering the duty cycle of the DC-DC converter to shift the PV operating voltage. If adjustment steers to an increase in power, the algorithm continues in that direction. Conversely, if the power level lowers, it reverses the direction of adjustment. This repeated technique helps the system to continuously "hover" around the Maximum Power Point (MPP), adapting to fluctuations in sunlight intensity and other environmental factors. From a practical standpoint, these duty cycle changes affect the input resistance of the buck converter, helping the PV array consistently operate near its peak power point. This ensures efficient energy transfer to the battery charge controller, which manages the EV battery's charging process.

Steps in the P&O Method

Monitoring Voltage and Current: The system continuously measures the PV array's voltage and current.

Perturbation and Observation: Small adjustments are made to the duty cycle, and the resulting changes in power output are observed.

Decision Making: Based on whether power increases or decreases, the algorithm determines whether to proceed with adjusting in the same direction or reverse course.

Implementing the P&O Algorithm in Simulink: In Simulink, the P&O algorithm can be implemented using basic blocks, avoiding the need for advanced coding (Patel, 2024; Mesentean *et al.*, 2011). The flow chart for the P and O algorithm is shown in Figure 7.2.

A unit delay block is used to track the previous power sample. A switch block handles the decision-making (if-else conditions).

The step size of the duty cycle adjustments is managed using a ΔD block, which allows fine-tuning of the algorithm's sensitivity.

To maintain safe and efficient operation, the duty cycle is constrained within a specified range (e.g., 0.4 to 0.6). This ensures that the system operates within safe limits while maximizing energy transfer to the EV battery.

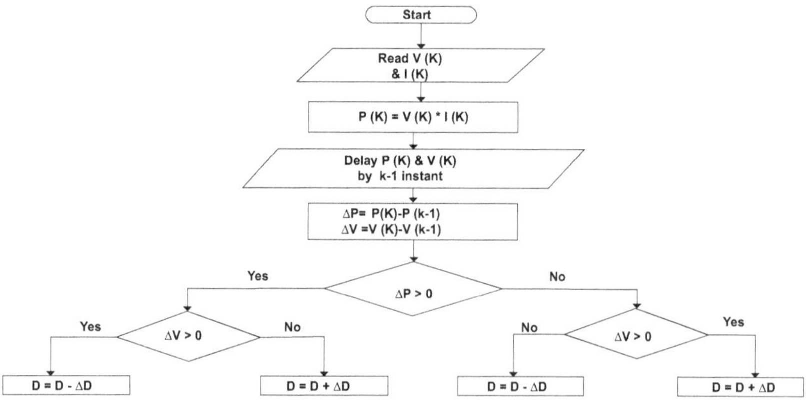

Figure 7.2 Flow chart of P&O algorithm.

Advantages of Using P&O in Solar Charging System

Efficiency with Low Complexity: Since P&O is computationally simple, it is ideal for systems where quick, low-power computations are preferred.

Responsive to Changes in Sunlight: The algorithm adapts quickly to fluctuating sunlight, maintaining efficient power flow even in partly cloudy or variable weather.

Direct Control of Input Resistance: The duty cycle adjustment affects the input resistance of the DC-DC converter, allowing precise control over the PV array's operating point.

7.9 LIMITATIONS AND FUTURE ENHANCEMENTS

One limitation of the P&O algorithm is that it can oscillate about the MPP, especially in rapidly changing weather conditions. Although this does not greatly impact small systems, larger setups may benefit from more advanced methods. Alternatives such as the Incremental Conductance method are often used where stability is critical, as they are less prone to oscillations. For applications where both stability and simplicity are required, some researchers are now experimenting with adaptive P&O or hybrid algorithms that combine P&O with other techniques (Goli & Shireen, 2014; Nunes *et al.*, 2015).

Overall, the P&O MPPT algorithm is an accessible and effective way to ensure solar PV panels provide maximum power to the EV charging system. With its simplicity, fast response, and compatibility with many small and medium setups, P&O remains a foundational technique for solar power applications. As renewable energy grows in importance, refinements in MPPT algorithms like P&O will continue to enhance the efficiency and reliability of solar-based EV charging systems, contributing to greener, more sustainable transportation.

7.10 SIMULATION RESULTS

A 250 W solar panel has been utilized for simulation purposes. The specifications of the solar panel (SPR-E20–250) are summarized in Table 7.1. The system is designed as a 4 kW photovoltaic (PV) setup, consisting of 16 panels arranged in a 4 × 4 configuration, with 4 panels connected in series and 4 such series connected in parallel. The simulation model is structured into distinct blocks representing the main components, designing an MPPT controller, developing a battery bank management system, implementing a bidirectional buck/boost converter, or creating a grid-tied inverter, and electric vehicles. These blocks function together to simulate the hybrid operation of

Table 7.1 Electrical Data Sheet of SPR-E20–250 (PV Panel)

Battery bank type and rating	*Lithium-ion, 48 V, 450 Ah*
EV battery rating	48 V, 30 Ah
Nominal power (P_{nom}) of PV panel	250 W
Total nos. of PV panels	16
PV panel configuration	4 × 4 (4 panels in one string and 4 strings)
Power tolerance	+ 5–0%
Average panel efficiency	20.4%
Rated voltage (V_{mpp})	30.2 V
Rated current (I_{mpp})	8.27 V
Open-circuit voltage (V_{oc})	36.6 V
Short-circuit current (I_{sc})	8.75 A
Maximum system voltage	1,000 V IEC and 60 V UL
Maximum series fuse	15A
Power temperature coefficient	– 0.35%/°C
Voltage temperature coefficient	– 176.6 mV/°C
Current temperature coefficient	2.6 mA/°C

the charging station, as illustrated in the accompanying figure of the charging station. The proposed simulation model for the project has been shown in Figure 7.3, comprising all the basic blocks.

Simulation Overview and Modes of Operation:

As per references (Singh *et al.*, 2021; Mouli *et al.*, 2016), various modes are:

Mode 1: The battery bank is charged using energy from the PV system.
Mode 2: Electric vehicles are charged directly by the PV system.
Mode 3: EVs are charged using grid power when the PV system cannot supply enough energy.
Mode 4: EVs are charged from the battery bank when neither the PV system nor the grid is available.
Mode 5: The PV system supplies power back to the grid.

7.11 WAVEFORMS OF THE 4KW PV SYSTEM'S MPPT/CHARGE CONTROLLER

Figure 7.4 (a) gives the capacity result by PV grid is proved at 1,000 W/m² and 25°C. The beginning ripple is on account of start of PV panels and PI controller. In Figure 7.4 (b), the PV current reaches a constant value of

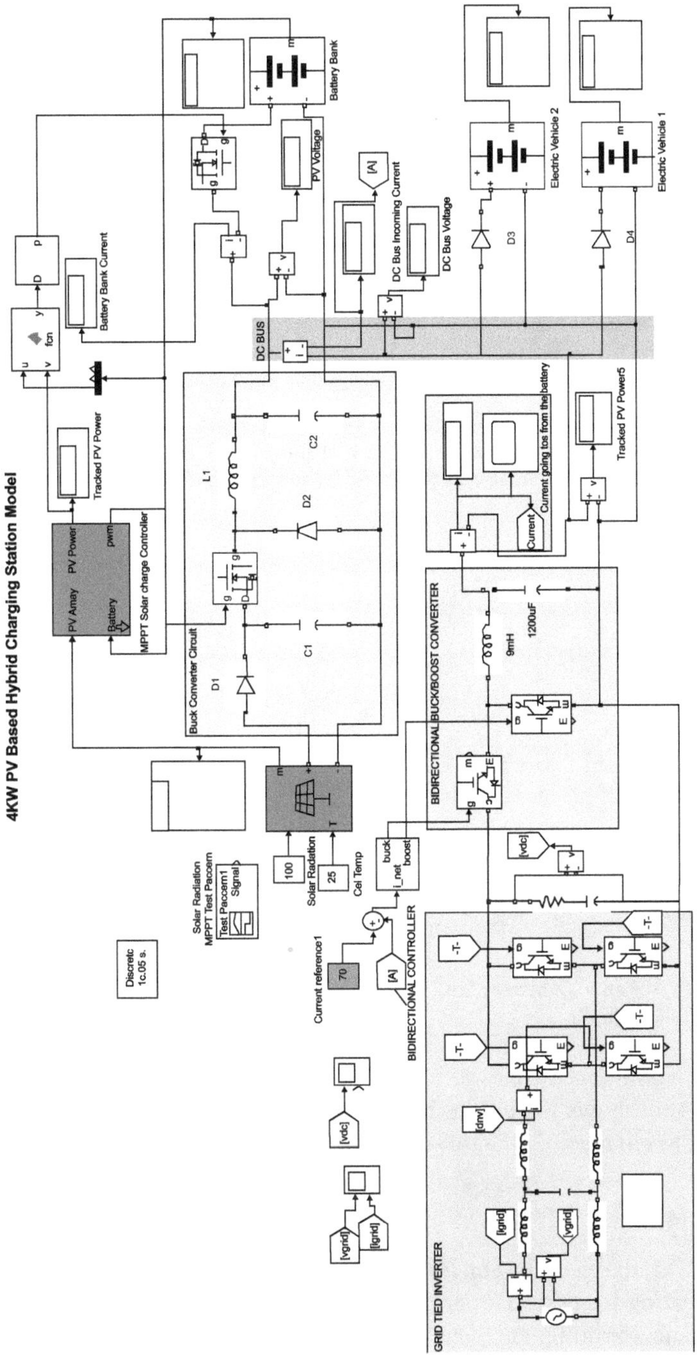

Figure 7.3 Hybrid charging station Simulink model.

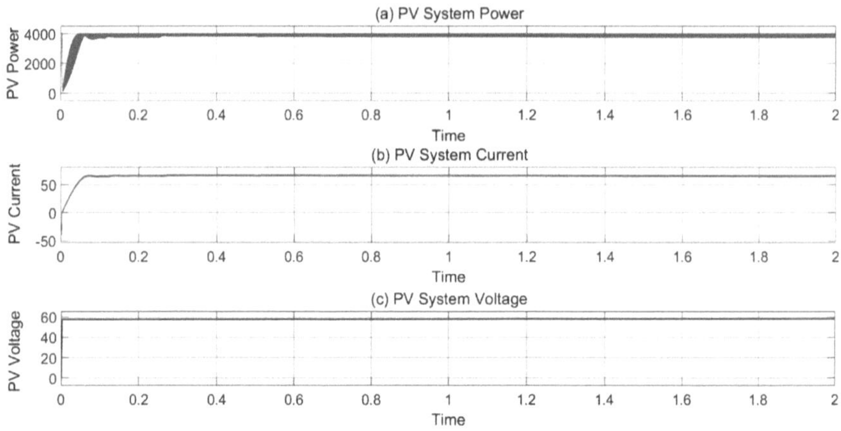

Figure 7.4 (a) Power changes tracked by the MPPT, (b) MPPT current (I), and (c) MPPT voltage (V) plotted against time at 25°C and 1,000 W/m² irradiance.

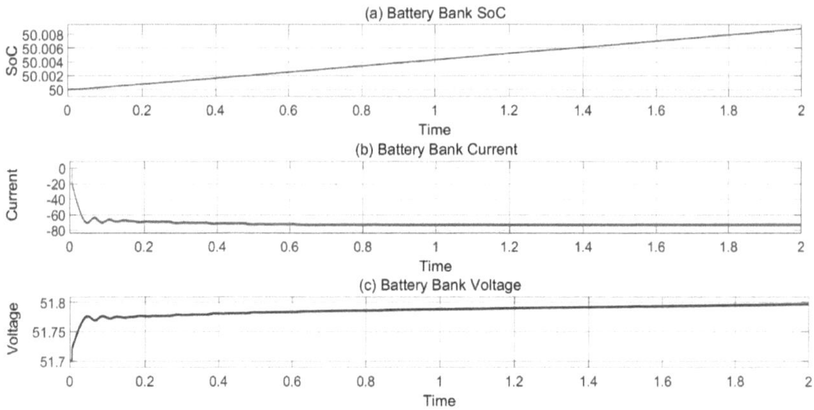

Figure 7.5 **(a)** Variation of battery SoC, (b) I (current) and (c) V (voltage) w.r.t time at 25°C, 1,000 Wb IR.

70 A for maximum capacity productivity following in position or time 0.5s. Figure 7.4 (c) shows the uninterrupted energized matter of 54 V across DC transport to charge EV's assault of 48 V.

7.11.1 Mode I

In Figure 7.5, the electric vehicles charging SoC, current, and physical ability are delineating in mode 1 operation when PV system charging the EVs as load now never-ending energized matter of 54 V across DC bus is used for charging the EVs and graph shows the increase in battery's state of charge when voltage and charging current is constant.

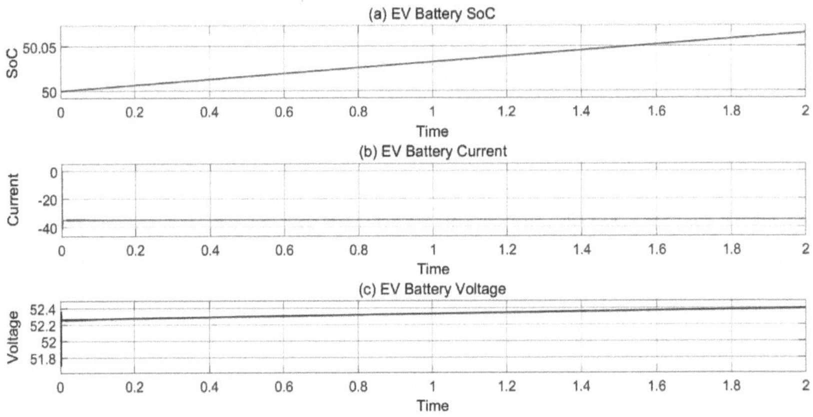

Figure 7.6 (a) Variation of EV's battery SoC, (b) current, and (c) voltage w.r.t time at 25°C, 1000 Wb IR.

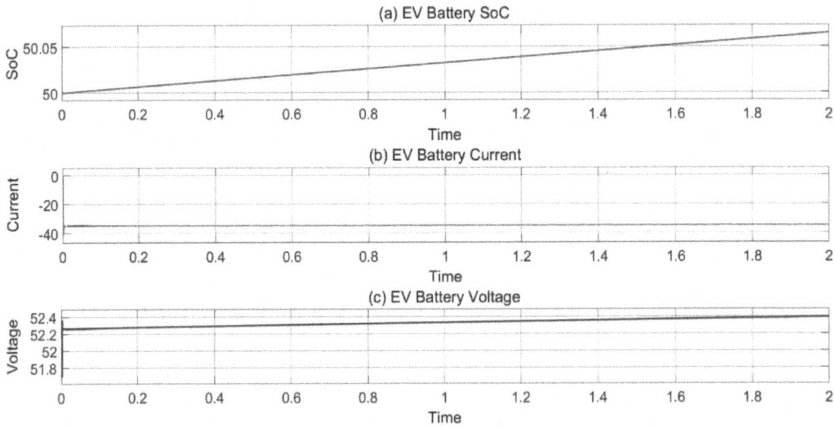

Figure 7.7 (a) Variation of EV's battery SoC, (b) I (current), and (c) V (voltage) w.r.t time at 25°C, 1000 Wb IR.

7.11.2 Mode 2

In Figure 7.6, the state of charge (SoC), current, and voltage of the EVs during Mode 2 operation are depicted.

7.11.3 Mode 3

Figure 7.7 represents the mode 3 operation when electric vehicles are charging all along midnight or moist season when PV capacity is insufficient to charge the EVs. In this mode, the EVs receive charging from the AC grid that has a bidirectional inverter and a bidirectional buck-boost converter.

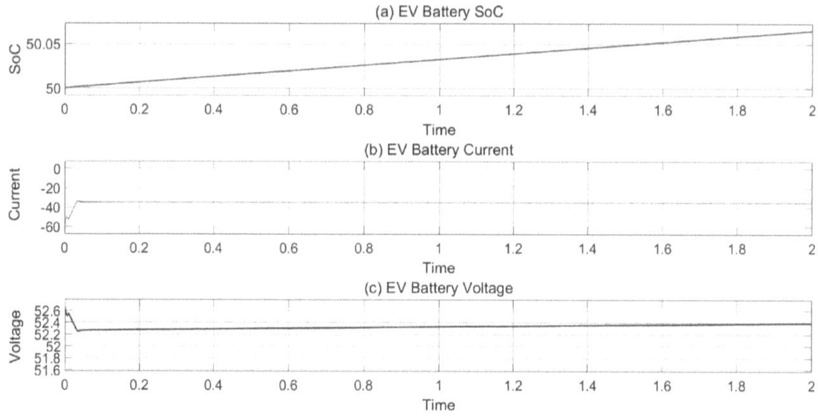

Figure 7.8 (a) Variation of EV's battery state of charge (b) current and (c) voltage in mode 4 operation.

7.11.4 Mode 4

In Figure 7.8, operation graphs of Mode 4 show the charging condition of EVs through battery bank. In this scenario, the battery bank acts as the primary energy source for charging EVs. This mode is particularly useful during extended power outages or periods when neither the grid nor the PV system is available. The simulation results demonstrate stable charging behavior, with the battery bank supplying the required current and voltage to the EVs.

7.11.5 Mode 5

When there are no active loads, such as fully charged batteries or idle EVs, the surplus energy generated by the PV system is fed back to the grid. This helps to stabilize grid load during peak hours and provides an additional source of income for the charging station operators. Figure 7.9 represents the graph of Voltage and current feed to the AC grid.

The results of the simulation for the five operational modes of the EV charging station were analyzed and confirmed using MATLAB® and PV system.

The simulation results display the usefulness of the hybrid charging station in ensuring uninterrupted power for EVs. The system can charge 10–12 EVs equipped with 48V, 30 Ah lithium-ion batteries. It also transfers excess solar energy to the grid, helping to reduce load demand. Furthermore, simulation results illustrate the operation of the charging station in various scenarios to a degree all the while at midnight or during the moist season when PV capacity is not enough, and all along long power cuts when the two together, the grid and PV method, are unavailable.

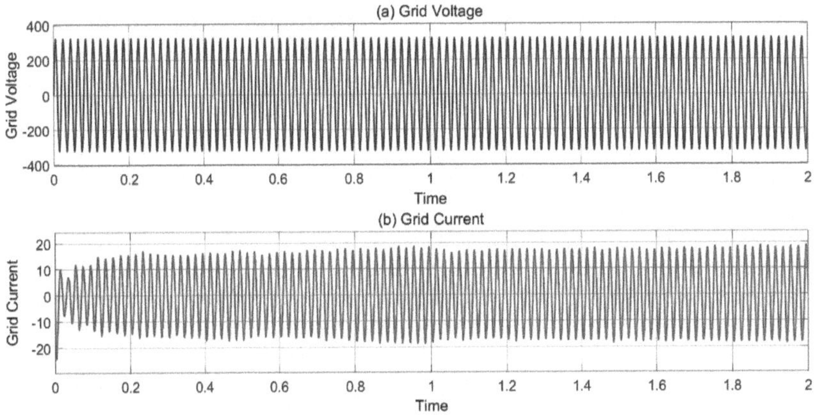

Figure 7.9 (a) Variation of voltage and (b) current graph w.r.t time at 25°C, 1000 Wb IR.

The simulation results validate the efficiency of the hybrid charging station in tackling challenges associated with grid stability and EV charging. Additionally, the system contributes to the advancement of sustainable transportation infrastructure and seamless renewable energy integration. Its ability to combine solar power with battery energy storage ensures a reliable power supply for EVs, representing a significant step toward reducing dependence on fossil fuels and alleviating grid congestion.

7.12 CONCLUSION

This chapter focuses on designing and simulating a solar-powered EV charging station, along with a hybrid charging controller. The system incorporates a three-stage charge controller, a buck converter, a grid-tied inverter, and an MPPT algorithm using the Perturb and Observe (P&O) method. Each component has been carefully analyzed and described, ensuring the design is both practical and replicable.

The charging station operates with a 4 kW photovoltaic (PV) system, which efficiently captures solar energy and regulates power delivery to charge 10–12 EVs equipped with 48 V, 30 Ah lithium-ion batteries. The system's design began with PV system software, which provided essential parameters. These parameters were then used to develop a simulation in Simulink, where outputs like voltage, current, power, and state of charge were evaluated and presented through detailed graphs.

While there is potential to refine aspects of the system, particularly the filtering process, the results demonstrate a highly functional and reliable solution. This chapter contributes to the growing field of renewable energy by integrating solar power into EV charging infrastructure. By addressing issues like grid stability and the increasing demand for EV charging, it supports the development of greener, more sustainable transportation systems.

REFERENCES

Aggarwal, S., Singh, A. K., Rathore, R. S., Bajaj, M., and Gupta, D. "Revolutionizing load management: A novel technique to diminish the impact of electric vehicle charging stations on the electricity grid," *Sustainable Energy Technologies and Assessments*, vol. 65, p. 103784, 2024, https://doi.org/10.1016/j.seta.2024.103784.

Bansal, P., Joshi, J., Hans, S., and Kumar, A. S. *Policies, Challenges and Advancements* (A. Kumar, A. Prasad, and G. Kumar, Eds.). John Wiley & Sons, 2024, pp. 79–103. https://doi.org/10.1002/9781394166831.ch5.

Chandra Mouli, G. R., Bauer, P., and Zeman, M. "System design for a solar powered electric vehicle charging station for workplaces," *Applied Energy*, vol. 168, pp. 434–443, Apr. 2016.

Geetha, P., and Usha, S. "Critical review and analysis of solar powered electric vehicle charging station," *International Journal of Renewable Energy Research*, vol. 12, no. 1, Mar. 2022.

Goli, P., and Shireen, W. "PV powered smart charging station for PHEVs," *Renewable Energy*, vol. 66, pp. 280–287, 2014.

Kumar, B. A., Jyothi, B., Singh, A. R., Bajaj, M., Rathore, R. S., and Berhanu, M. "A novel strategy towards efficient and reliable electric vehicle charging for the realisation of a true sustainable transportation landscape," *Scientific Reports*, vol. 14, no. 1, p. 3261, 2024, https://doi.org/10.1038/s41598-024-53214-w.

Mesentean, S., Feucht, W., Mittnacht, A., and Frank, H. "Scheduling methods for smart charging of electric bikes from a grid-connected photovoltaic-system," in *Proceedings—UKSim 5th European Modelling Symposium on Computer Modelling and Simulation (EMS 2011)*, pp. 299–304, Nov. 2011, https://doi.org/10.1109/EMS.2011.88.

Nunes, P., Farias, T., and Brito, M. "Enabling solar electricity with electric vehicles smart charging," *Energy*, vol. 87, Jun. 2015, https://doi.org/10.1016/j.energy.2015.04.044.

Patel, M. K. "Energy management strategies of grid connected renewable source for EV charging station," *International Journal of Scientific Research in Science, Engineering and Technology*, vol. 11, no. 2, pp. 473–482, Apr. 2024, https://doi.org/10.32628/IJSRSET2411270.

Pinjari, T. S., Shinde, S., Salunkhe, R., Gadhave, S., and Bansode, S. "Solar charging station for electric vehicles," *International Journal of Advance Research and Engineering*, vol. 2, no. 1, 2016.

Ramakrishnan, V., Vishnuram, P., Yang, T., Bajaj, M., Rathore, R. S., and Zaitsev, I. "Design and implementation of a high misalignment-tolerance wireless charger for an electric vehicle with control of the constant current/voltage charging," *Scientific Reports*, vol. 14, no. 1, p. 13165, 2024, https://doi.org/10.1038/s41598-024-63952-6.

Ravindran, M. A., Nallathambi, K., Vishnuram, P., Rathore, R. S., Bajaj, M., Rida, I., and Alkhayyat, A. "A novel technological review on fast charging infrastructure for electrical vehicles: Challenges, solutions, and future research directions," *Alexandria Engineering Journal*, vol. 82, pp. 260–290, 2023, https://doi.org/10.1016/j.aej.2023.10.009.

Shafiq, S. I., Rehman, A. U., Elbarbary, Z. M. S., Kotb, H., Selim, A., and Kamel, S. "Integration of solar-based charging station in power distribution network and charging scheduling of EVs," *Frontiers in Energy Research*, vol. 11, Jan. 12, 2023.

Shariff, S. M., Alam, M. S., Ahmad, F., Rafat, Y., Asghar, M. S. J., and Khan, S. "System design and realization of a solar-powered electric vehicle charging station," *IEEE Systems Journal*, vol. 14, no. 2, Jun. 2020.

Singh, A. R., Vishnuram, P., Alagarsamy, S., Bajaj, M., Blazek, V., Damaj, I.,. . . Othman, K. M., "Electric vehicle charging technologies, infrastructure expansion, grid integration strategies, and their role in promoting sustainable e-mobility," *Alexandria Engineering Journal*, vol. 105, pp. 300–330, 2024a, https://doi.org/10.1016/j.aej.2024.06.093.

Singh, P. P., Das, S., Wen, F., Palu, I., Singh, A. K., and Thakur, P. "Multi-objective planning of electric vehicles charging in distribution system considering priority-based vehicle-to-grid scheduling," *Swarm and Evolutionary Computation*, vol. 77, p. 101234, 2023, http://dx.doi.org/10.1016/j.swevo.2023.101234.

Singh, S. S., Sekhar, R., Shaik, S., and Ghosh, A. "Design and analysis of a solar-powered electric vehicle charging station for Indian cities," *World Electric Vehicle Journal*, vol. 12, pp. 1–18, Aug. 2021, https://doi.org/10.3390/wevj12030132.

Singh, V. P., and Kumar, A. "Techno-economic and future aspects of the HEV-EV-FCV: Decarbonization, digitalization and sustainability," in *Energy Efficient Vehicles Technologies and Challenges*, V. P. Singh, A. Kumar, C. S. Meena, and G. Dwivedi, Eds. Boca Raton, FL: CRC Press, 2024, pp. 212–231, https://doi.org/10.1201/9781003464556–11.

Singh, V. P., Kumar, A., Meena, C. S., and Dwivedi, G., Ed. *Energy Efficient Vehicles: Technologies and Challenges*, CRC Press, Boca Raton, FL, 2024b, https://doi.org/10.1201/9781003464556. ISBN: 9781032548111.

Singla, P., Boora, S., Singhal, P., Mittal, N., Mittal, V., and Gared, F. "Design and simulation of 4 kW solar power-based hybrid EV charging station," *Scientific Reports*, vol. 14, p. 7336, 2024, https://doi.org/10.1038/s41598-024-56833-5.

Sivasami, T. S. K., and Khan, S. "Design and implementation of solar charging station for electric vehicles," *IJMRD*, vol. 7, no. 2, Oct. 2023.

Tan, R. H. G., Er, C. K., and Solanki, S. G. "Modeling of photovoltaic MPPT lead acid battery charge controller for standalone system applications," *E3S Web of Conferences*, vol. 182, p. 03005, 2020.

Upadhyay, R. K., Singh, V. P., and Kumar, A. "Sustainable transportation: Policy, planning and implementation," in *Energy Efficient Vehicles Technologies and Challenges*, V. P. Singh, A. Kumar, C. S. Meena, and G. Dwivedi, Eds. Boca Raton, FL: CRC Press, 2024, pp. 74–95, https://doi.org/10.1201/9781003464556-5.

Chapter 8

Performance Assessment of Five-Level Inverter for Solar Photovoltaic Systems

Ashish Srivastava, Subho Upadhyay, Anurag Chauhan, and Anurag Tripathi

8.1 INTRODUCTION

The introduction of multilevel inverter technology marks a significant milestone in power electronics. Multilevel inverters generate an output waveform with several voltage levels that help in reducing the harmonic distortion along with improving the quality of power [1, 2]. Multilevel topologies have gained immense attention in choosing a balance between performance, complexity, and cost. Among different MLIs, the five-level configuration is very suitable for solar applications as it produces excellent power quality with lower switching loss and higher efficiency when compared to two-level converters [3].

Cutting-edge five-level inverters are needed to adapt to and conform to the specific needs of solar PV systems that have the peculiarities of renewable energy sources. Solar systems generate power that may be inconsistent due to variable sunlight and temperature levels. This inconsistency in power quality would prove painful, especially during the connection of the solar PV system to the grid and sensitive loads. The five-level inverters solve this by providing a very clean output sine wave with minimal harmonic distortion and minimizing electromagnetic interference.

Next to power quality, higher system efficiency with the reduction of switching loss will be of benefit from the five-level inverter. With the old traditional two-level inverters, a high-frequency switching technique is continuously being engaged, whose application causes a lot of energy loss every moment of switching [4]. On the other hand, switching in five-level case disperses voltage over several switches, therefore reducing the burden on single malfunctions and lowering losses. This is of utmost importance for solar systems, where maximum energy conversion efficiency contributes to functionality and cost-effectiveness.

Another application of the usage of five-level inverters is in solar PV systems, where to come back on the commercial end, the large trading utilities are also in need for that matter. Since there will be a good demand for high-power inverters with effective operation for solar farms and microgrids,

DOI: 10.1201/9781003634737-8

five-level inverters will be in an excellent position in regard to such demands due to flexibility, reliability, and compatibility with other renewable energy systems [5].

Besides solar applications, five-level inverters can find technologies suitable for hybrid systems for wind, biomass, and energy storage. The systems demand advanced power electronics to guarantee their smooth operation and capability of energy utilization, making five-level inverters a good candidate. In the future, consideration will be given to ongoing research aimed at making five-level inverters even performance-oriented and cost-effective. There are new trends: the use of artificial intelligence and machine learning for real-time control, along with research into new circuit designs and materials to further improve the efficiency and reliability. As these advances go on, five-level inverters will have an important future with the integration of renewable energy systems [6].

8.1.1 Multilevel Inverters

Multilevel inverters (MLIs) exhibit numerous advantages, including enhanced power quality, reduced harmonic distortion, decreased switching losses, improved electromagnetic interference, and overall performance enhancement. Consequently, the adoption of multilevel inverters has surged in recent years within industrial applications. The principal topologies of multilevel inverters include the following [7, 8]:

- Diode Clamped MLI
- Flying Capacitor MLI
- Cascaded H-Bridge MLI

Illustration of the output of a five-Level Multilevel Inverter (MLI) compared to a reference sine wave [9] is shown in Figure 8.1. The dashed line represents the ideal AC waveform we aim to achieve, while the solid blue line shows the five-level stepped waveform generated by the MLI, consisting of five distinct voltage levels. The inverter generates these levels (e.g., –V, –V/2, 0, V/2, V) using a combination of switches and capacitors or separate DC source. By switching between these voltage levels, the inverter produces a stepped approximation of the desired sine wave, with more levels resulting in a waveform closer to the sine wave. The switching process is controlled by methods like Sinusoidal Pulse Width Modulation (SPWM) or other modulation methods to ensure the output closely follows the reference sine wave [10].

8.1.2 Switching Techniques for Multilevel Inverters

Switching in a single-phase multilevel inverter plays an important part in deciding the quality and efficiency of the output waveform. The primary

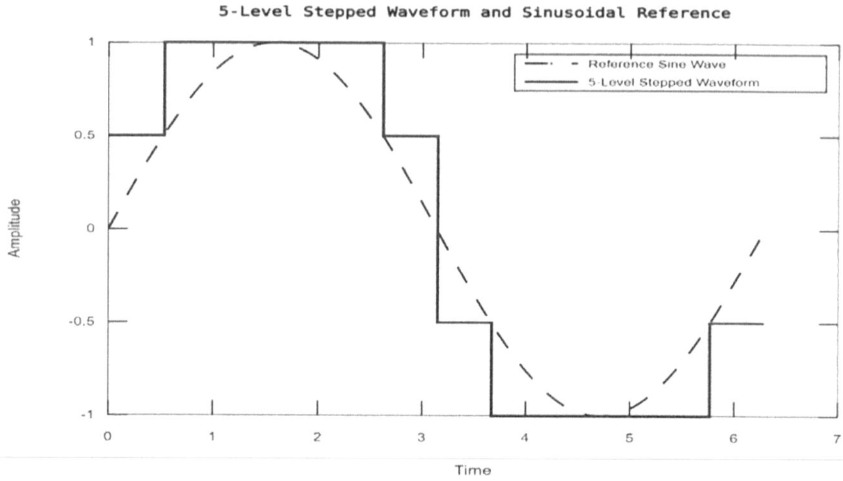

Figure 8.1 Stepped waveform of MLI.

requirement of switching in MLIs is to achieve a high-quality AC output that closely approximates a sinusoidal waveform. This is accomplished by controlling the power electronic switches to generate multiple voltage levels from a DC input. In a single-phase MLI, the switches are operated in a manner that sequentially connects the DC voltage sources or capacitors that connect to the output, creating a stepped waveform that mimics a sine wave. This stepped waveform has greatly lowered harmonic distortion in comparison with traditional two-level inverters, which is one of the main reasons for employing MLIs in applications like renewable energy systems, especially solar photovoltaic setups.

By raising the number of voltage steps through proper switching, the inverter can produce an output waveform with lower total harmonic distortion (THD), which improves the power quality. Additionally, the reduced harmonic content in the output minimizes the stress on the connected loads, enhancing their lifespan and performance. Another advantage is the reduction in switching losses. Since the voltage transitions between adjacent levels are smaller, the energy lost during these transitions is significantly lower compared to traditional inverters, resulting in higher overall efficiency. Furthermore, the lower voltage stress on each individual switch allows for the use of components with lower ratings, which can reduce the cost and complexity of the inverter design.

One of the primary challenges is the complexity of the control algorithms required to manage the multiple switches effectively [11, 12]. With a number of voltage steps, it becomes harder to manage the control logic such as Sinusoidal Pulse Width Modulation (SPWM), Space Vector Pulse Width Modulation (SVPWM), or Selective Harmonic Elimination (SHE). These algorithms

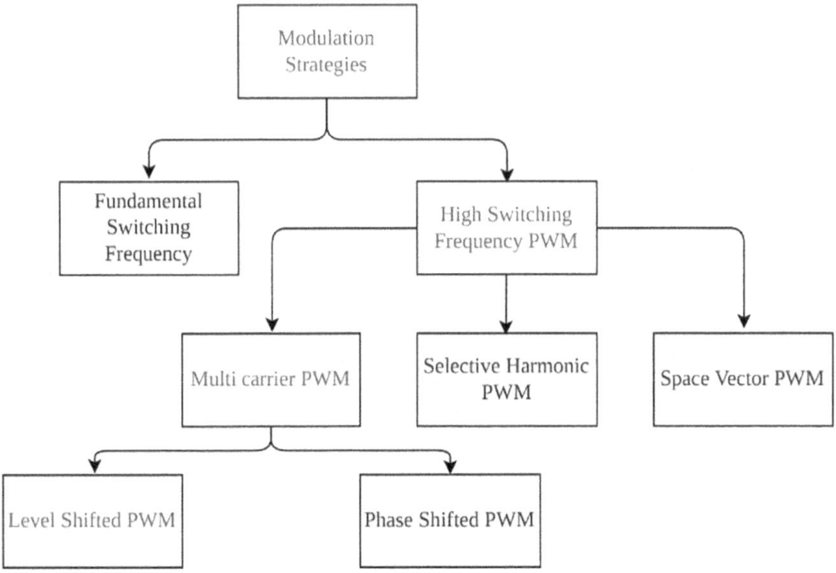

Figure 8.2 Different switching techniques of MLI.

demand precise timing and coordination, which can complicate the design and increase the computational burden on the control system. Moreover, the need for more switches and additional components such as capacitors or separate DC sources in higher-level inverters can lead to increased system size, cost, and potential reliability concerns. Different switching techniques of MLI are depicted in Figure 8.2.

A comparison of different switching techniques of MLI is given in Table 8.1. Switching methods, that is, Sinusoidal Pulse Width Modulation (SPWM), Space Vector Modulation (SVM), Selective Harmonic Elimination (SHE), Phase Disposition PWM (PDPWM), Phase Opposition Disposition PWM (PODP), and Neutral Point Clamped PWM (NPC PWM), have been considered under the present study. A brief discussion on advantages, disadvantages, and applications of each modulation technique is summarized in Table 8.1. Among various methods, LSPWM has been considered for the performance investigation of the five-level MLI. This switching method reduces the waveform deviation, power losses, and THD by spreading out the switching evenly across levels. The control logic of LSPWM offers easier operation as compared to SVM and SHEPWM. Further, the voltage and current waveforms of the five-level inverter are traced in the chapter. Furthermore, performance of the MLI is verified for different operating frequencies. Finally, THD analysis of the inverter is computed by using the FFT toolbox in MATLAB®.

Table 8.1 Different Modulation Techniques for Multilevel Inverter [10]

S. No.	*Modulation Techniques*	*Description*	*Advantages*	*Disadvantages*	*Applications*
1	Sinusoidal Pulse Width Modulation (SPWM)	It uses sinusoidal reference and triangular carrier waves to generate gate signals.	• Simple and easy to implement • Low harmonic distortion at lower modulation index • Good voltage control	• Limited modulation index range • High switching losses due to higher carrier frequency • Reduced efficiency at higher levels	Motor drives, renewable energy systems [13]
2	Space Vector Modulation (SVM)	Uses vector representation of reference voltage to optimize switching.	• Increased utilization of the DC bus system • Lower total harmonic distortion • Suitable for high-performance of MLI	• Complex implementation • Requires advanced computation of resources • Difficult for systems with many levels	Performance enhancement of motor control, aerospace [14]
3	Selective Harmonic Elimination (SHE)	Eliminates specific lower-order harmonics by solving nonlinear equations.	• Very low harmonic distortion • Optimal efficiency. • Flexible harmonic suppression capability	• Complex mathematical computation • Limited to specific harmonic elimination • Requires offline computation	Medium- and high-power industrial systems [15]
4	Phase Disposition PWM (PDPWM)	All carriers are in-phase for modulation.	• Simple to implement • Suitable for multilevel inverters with fewer levels • Balances output voltage effectively	• Higher harmonic content compared to advanced techniques • Limited efficiency improvement at higher levels	Renewable energy systems, medium power drives [16, 17]

(Continued)

Table 8.1 (Continued)

S. No.	*Modulation Techniques*	*Description*	*Advantages*	*Disadvantages*	*Applications*
5	Phase Opposition Disposition PWM (PODP)	Carriers above and below zero line are in opposite phases.	• Balanced harmonic performance • Suitable for higher number of levels • Ensures voltage balancing	• Higher switching losses compared to optimized techniques • Implementation becomes complex for very high levels	Medium- to high-power converters, HVDC systems [18]
6	Neutral Point Clamped PWM (NPC PWM)	Modulation specific for neutral-point clamped inverters.	• Efficient DC link utilization • Balances DC link capacitor voltages • Suitable for multilevel topology	• Increased switching complexity • Sensitive to DC link voltage imbalance • Requires specific topology support	High voltage industrial systems, power grids [19]
7	Level-Shifted PWM (LSPWM)	Carrier signals are shifted in level with respect to each other.	• Simple implementation • Reduces total harmonic distortion • Effective for multilevel inverters with fewer levels	• Limited scalability to very high levels • Moderate switching losses	Medium power applications, residential inverters [20]

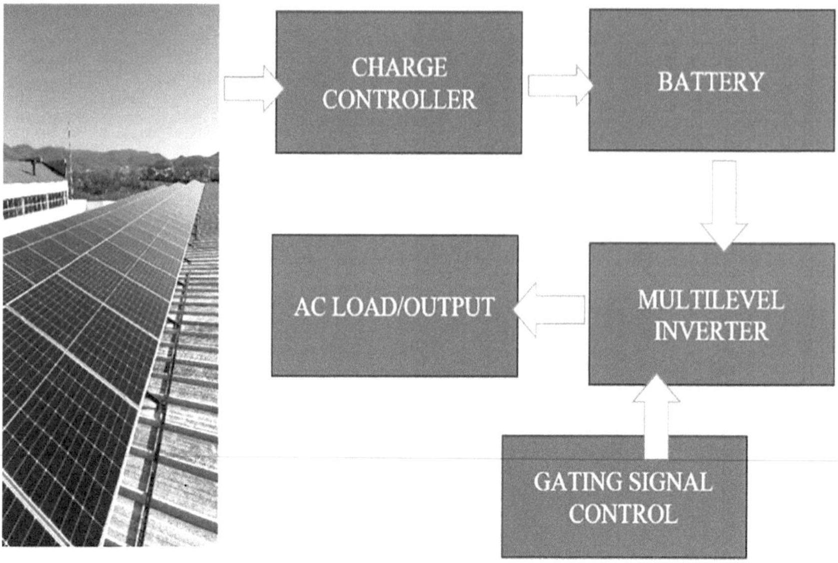

Figure 8.3 Schematic of solar PV-based multilevel inverter.

8.2 DESCRIPTION OF THE PROPOSED SYSTEM

A schematic representation of a solar PV-based multilevel inverter is shown in Figure 8.3. The straightforward and reliable method for the conversion of DC into AC voltage requires a multilevel inverter and multiple DC sources. At first step, solar panels take the sunlight and turn it into electrical energy. Further, the process begins with this energy going to a charge controller. This device makes sure the voltage and current are just right to stop the batteries from getting overcharged. After that, the energy is stored in a battery. Furthermore, the stored energy is supplied to a multilevel inverter to convert DC into alternating current. Finally, the converted AC power is supplied to the AC Loads.

A solar photovoltaic system converts the raw energy from the sun into electrical power. The effect of an increase in solar radiation also increases the photovoltaic current and increases the output power, while an increase in temperature reduces the output voltage and hence reduces the overall power. The output power of SPV depends on the number of solar modules, fill factor, open circuit voltage, and short circuit current of a photovoltaic module. Here, Nm is the total number of solar modules, V_{OC} is open circuit voltage, I_{sc} is short circuit current (A), FF is form factor of the solar panel, and PSPV is the output power of the solar photovoltaic system (kW) [7].

$$P_{SPV} = \frac{N_m \times V_{oc} \times I_{SC} \times FF(t)}{1000} \quad (8.1)$$

Short circuit current and open circuit voltage are calculated using Equations 8.2 and 8.3, respectively. The cell temperature of a photovoltaic cell is related to ambient temperature and solar radiation and is given in the following equation. Here, I_{sc} is short circuit current (A), I_{sc}, stc is short circuit current at standard test conditions, K_i is temperature coefficient of current (A/^{0}C), Tc is cell temperature and G is solar radiation (W/m^2), Tais ambient temperature and NOCT is nominal operating cell temperature (^{o}C).

$$I_{SC}(t) = \left\{I_{sc,stc} + K_i\left[T_c(t) - 25^o\right]\right\}\frac{G}{1000} \quad (8.2)$$

$$T_C(t) = T_a(t) + \frac{NOCT - 20^o}{800} \times G \quad (8.3)$$

Similarly, open circuit voltage can be estimated as shown in the following equation with Voc,stc as open circuit voltage at standard test conditions (V) and K_v as temperature coefficient of voltage (V/^{o}C).

$$V_{oc}(t) = V_{oc,\,stc} - K_v \times T_c(t) \quad (8.4)$$

8.2.1 Level-Shifted PWM: Principles and Operation

In Level-Shifted PWM (LSPWM), the number of carrier signals needed for a multilevel inverter is N–1. The signals should have equal frequency and equal amplitude. All carrier signals are phase shifted with respect to each other taking zero reference into consideration. Carrier signals are compared with the modulating signal, and a pulse is generated at the instants when the amplitude of the modulating signal is greater than the carrier signal. Under the sinusoidal pulse width modulation, the reference is sine wave as shown in the simulation model. After a comparison is done, the output signals of the comparator are transferred to the IGBT switches.

The working of the LSPWM method is depicted in Figure 8.4. The sinusoidal waveform is first compared with the carrier signals at each instant to produce the gate signals. When the amplitude of the reference waveform is greater than the carrier signals, a positive pulse of fixed magnitude is generated for a finite time. However, no pulse is produced under the condition when the amplitude of carrier wave is greater than sine wave.

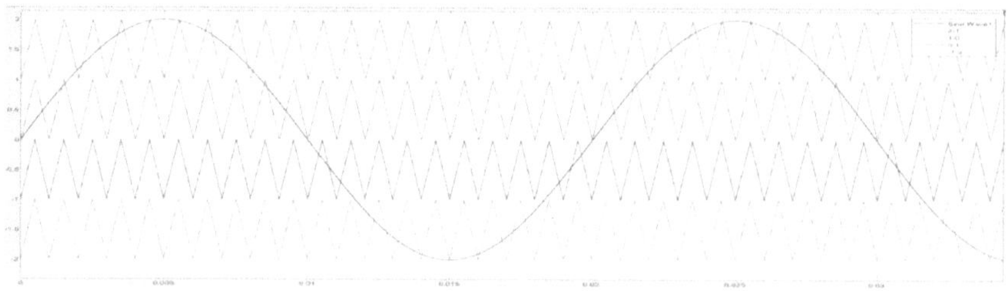

Figure 8.4 Carrier waveforms and sine waveforms for modulation of MLI.

8.3 MLI TOPOLOGIES

8.3.1 Diode-Clamped MLI

A diode-clamped multilevel inverter (n-level) generates “n” distinct voltage levels using diodes for voltage clamping. It is characterized by its output voltage levels and the components required for its operation. Each leg of the inverter consists of:

$$\text{Number of capacitors} = (n-1) \tag{8.5}$$

$$\text{Number of switches} = 2(n-1) \tag{8.6}$$

$$\text{Number of clamping diodes} = (n-1)(n-2) \tag{8.7}$$

According to Equations 5–7, the corresponding specifications for a five-level diode-clamped multilevel inverter would be as follows:

The number of capacitors that are on the DC-link bus is (5 – 1) = 4.
Number of switches required in each leg = 2 × (5 – 1) = 8.
The number of clamping diodes is (5 – 1) (5 – 2) = 12

As per the calculations made, a schematic of the proposed five-level diode clamped inverter is depicted in Figure 8.5, which shows that five-level inverter needs 8 switches in each leg, 12 clamping diodes, and 4 capacitors. Different capacitors (C_1, C_2, C_3, and C_4) are connected in series to have the same voltage level. The center point between the capacitors is considered the neutral point. The circuit uses clamping diodes (D_1, D_2, D_3, etc.) to restrict the voltage across each switching device to one capacitor voltage level, ensuring the safe operation of switches. The working principle is based on the controlled switching of IGBT (S_{11} to S_{14} and S_{11}’ to S_{14}’) to produce the desired multilevel output.

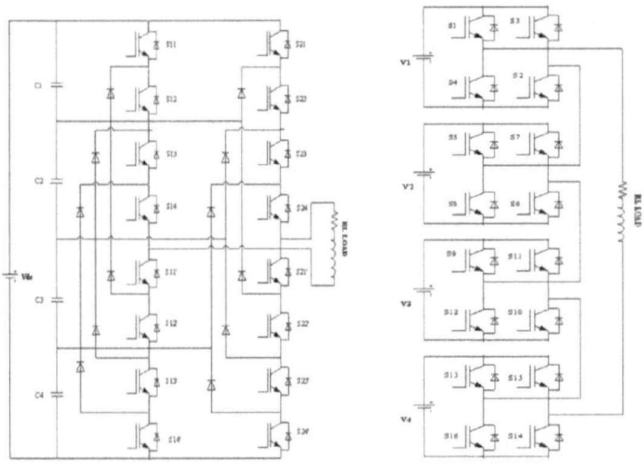

(i) Diode clamped inverter (ii) CHB inverter

Figure 8.5 Configuration of five-level inverter.

When all the upper switches S_{11} to S_{14} are turned "On," a voltage level V_{dc} will be obtained across the load. While three upper switches, S_{12}, S_{13}, and S_{14}, along with one lower switch S_{11}', are conducting, a voltage level of $3V_{dc}/4$ is obtained across the load. When two upper switches S_{13} and S_{14} with two lower switches S_{11}' and S_{12}' are turned on, a voltage step of $V_{dc}/2$ voltage is achieved across the load. When all the lower switches S_{11}' to S_{14}' are turned on, zero voltage will be attained across the load [11].

8.3.2 Cascaded H-Bridge Multilevel Inverter

Cascaded configuration of MLI comprises the series combination of H-bridge units. The desired voltage is achieved from various distinct DC sources, which can be collected from solar modules, fuel cells, batteries, etc. This type of configuration does not need any capacitor for voltage balancing or diode for voltage clamping.

For an n-level inverter, the following is valid:

$$\text{Number of H} - \text{bridge cells} = n - 1 \tag{8.8}$$

Every H-bridge cell needs four switches.

$$\text{The total number of switches} = 4 \times (n - 1) \tag{8.9}$$

According to Equations 11–12, five-level cascaded H-bridge multilevel inverter:

$$\text{Number of H} - \text{bridge cells} = (5 - 1) = 4$$

Since there are 4 switches per H-bridge, total no. of switches = 4 × 4 = 16. A five-level cascaded H-bridge multilevel inverter requires 4 H-bridge cells, 16 switches, and 4 isolated DC sources. Capacitor and clamping diodes do not exist in this topology. The circuit shown in Figure 8.5 is a Cascaded H-Bridge Multilevel Inverter (CHB-MLI). It produces a strong AC output using several DC voltage sources. The design features 2 H-bridges, each connected to its own DC source and controlled by 16 switches (either IGBTs or MOSFETs). The four switches on each H-bridge are turned on and off to produce the three voltage levels of $+V_{dc}$, 0 V_{dc}, and $-V_{dc}$. The inverter produces a stepped output with numerous voltage levels by connecting many H-bridges.

8.4 RESULTS AND DISCUSSIONS

In this section, performance parameters of the five-level MLI have been assessed using MATLAB® Simulink. At first, output waveforms of the inverter have been tested to verify the total voltage steps. Further, shape of the current waveform through the load is traced to confirm the deviation from the ideal sinusoidal waveform. Finally, overall THD and percentage share of lower harmonics in the THD of the load voltage are computed.

8.4.1 Waveforms of Voltage and Current

Two MLI topologies of five-level inverter, that is, diode clamped and CHB inverter, have been modeled in MATLAB®. Further, these two configurations are simulated to supply the load demand of 50Ω+50mH. Waveform results of diode clamped inverter across and through the load have been investigated, as depicted in Figure 8.6. It is determined that all mandatory voltage

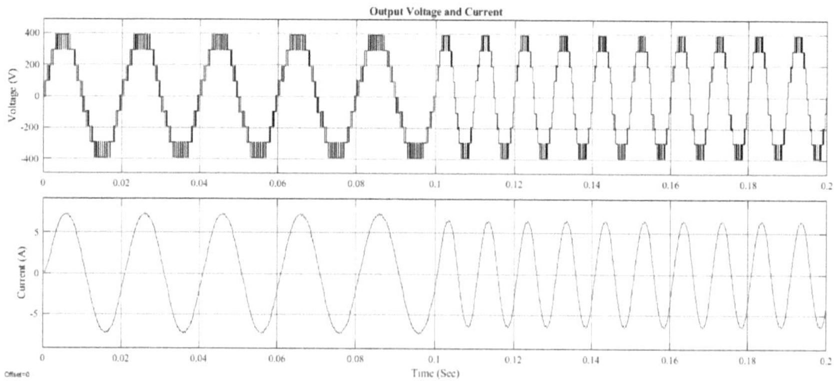

Figure 8.6 Waveform results of five-level diode clamped inverter.

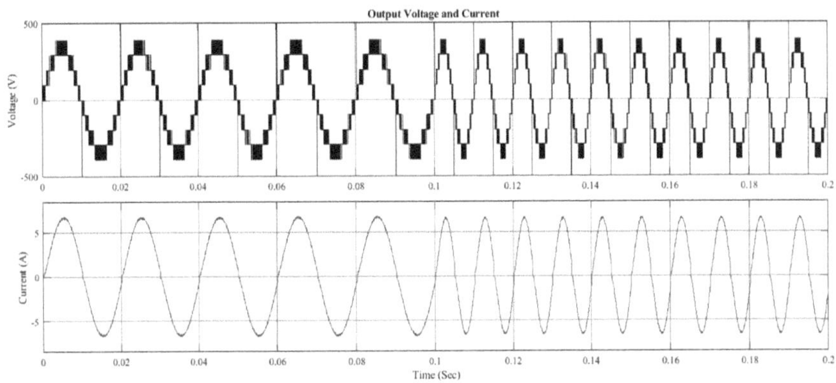

Figure 8.7 Waveform results of CHB inverter.

steps are achieved in the waveform. Also, shape of the current waveform is found to be sinusoidal. Also, simulation of CHB inverter has been performed to attain five levels in the load voltage as illustrated in Figure 8.7. It is perceived that CHB configuration offers all basic voltage steps in the waveform. Further, current waveform is more sinusoidal than diode clamped.

8.4.2 Effect of Frequency

Further, supply frequency is modified during the simulation from 50 Hz to 100 Hz to test the performance of the MLI configurations under dynamic change in frequency, as illustrated in Figure 8.6 and Figure 8.7, respectively. Frequency of 50 Hz is maintained from 0 to 0.1 sec; further, frequency is changed to 100 Hz at t = 0.1 sec; and it is kept till t = 0.2 sec. It is determined that the performance of both MLI configurations is consistent under the dynamic variation in frequency.

8.4.3 THD Spectrum

Finally, harmonics in the load voltage of five-level diode-clamped inverter are computed using MATLAB®. Percentage THD in the output voltage is calculated as 15.68% as presented in Figure 8.8. It is concluded that the 3rd, 9th, and 13th order of harmonics are below 1% in the output voltage. While 5th, 7th, 11th, and other higher-order harmonics are found to be more than 1% in the load voltage.

Furthermore, the THD spectrum of the considered five-level CHB inverter has been analyzed as depicted in Figure 8.9. It is inferred that THD of 13.60% is computed in the load voltage using the FFT toolbox of MATLAB®. All lower-order harmonics (3rd, 5th, 7th, 9th) in the load voltage are estimated

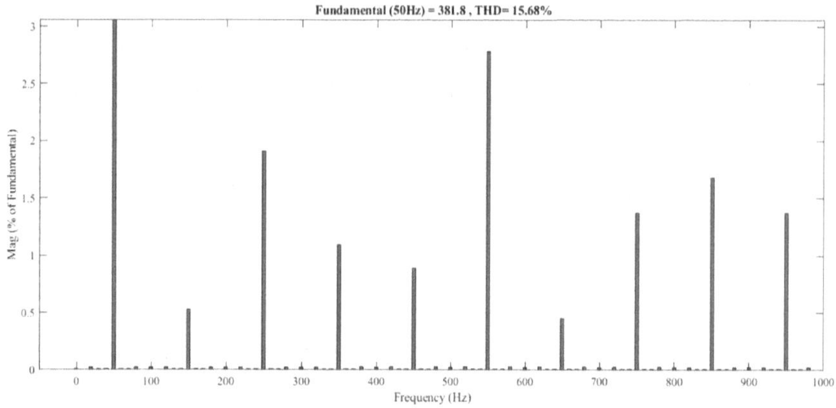

Figure 8.8 THD computation in the load voltage of five-level diode clamped inverter.

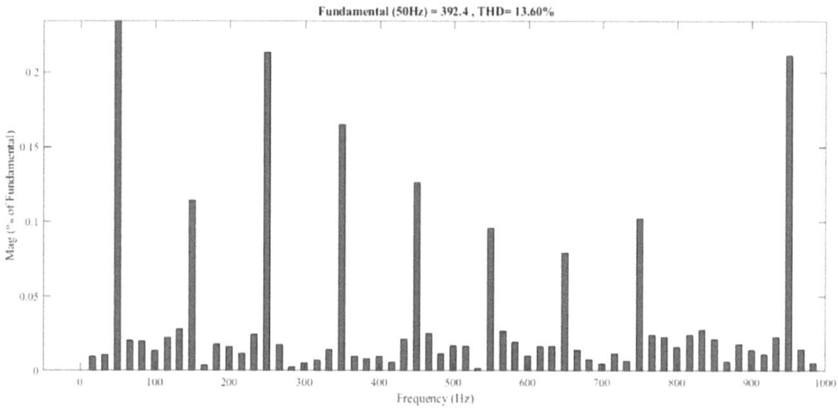

Figure 8.9 THD spectrum of load voltage for five-level CHB inverter.

and found to be below 1%. Power quality of the load voltage in the CHB inverter is found to be superior as compared to the diode clamped inverter.

8.5 CONCLUSIONS

In the present chapter, a performance study of a five-level inverter has been carried out using the LSPWM technique. Two different MLI topologies, that is, diode clamped and cascaded H-bridge inverters are considered in this chapter. Further, waveform results of MLI output voltage and current are analyzed using MATLAB® Simulink. It is perceived that all voltage steps are achieved in the load voltage. Also, the current waveform is found to be more sinusoidal in the case of the CHB inverter.

Furthermore, the performance of the considered MLI topology remains consistent for the change in supply frequency, that is, 50 Hz to 100 Hz. Also, THD values of the MLI topology are computed as 15.68% and 13.60% for diode clamped and CHB configurations, respectively. The considered configuration is suitable for medium voltage and high-power applications, such as solar photovoltaic and motor drives. Future studies of multilevel inverters may be carried out to minimize the number of switches and improve the switching algorithms, operation in smart grid, and testing in real time.

REFERENCES

[1] Lai, J.-S., & Peng, F. Z. (1996). Multilevel converters: A new breed of power converters. IEEE Transactions on Industry Applications, 32(3), 509–517.

[2] Panda, K. P., Bana, P. R., & Panda, G. (2020). FPA optimized selective harmonic elimination in symmetric–asymmetric reduced switch cascaded multilevel inverter. IEEE Transactions on Industry Applications, 56(3), 2862–2870.

[3] Babu, T. R., Kumar, K. S., & Saravanan, C. (2020). Performance analysis of five-level inverter topologies for solar PV applications. International Journal of Power Electronics and Drive Systems, 11(3), 1482–1492.

[4] Kouro, S., Malinowski, M., Gopakumar, K., Pou, J., Franquelo, L. G., Wu, B., Rodriguez, J., Perez, M. A., & Leon, J. I. (2010). Recent advances and industrial applications of multilevel converters. IEEE Transactions on Industrial Electronics, 57(8), 2553–2580.

[5] Carrasco, J. M., Franquelo, L. G., Bialasiewicz, J. T., Galván, E., Guisado, R. C. P., Prats, M. Á. M., León, J. I., & Moreno-Alfonso, N. (2006). Power-electronic systems for the grid integration of renewable energy sources: A survey. IEEE Transactions on Industrial Electronics, 53(4), 1002–1016.

[6] Azeem, A., Ali, M. U., Sattar, A., Naqvi, I. H., Rana, A., & Mehmood, C. A. (2021). A comprehensive review on multilevel inverters for PV-based renewable energy applications. Energies, 14(6), 1624.

[7] Upadhyay, S., & Sharma, M. P. (2016). Selection of a suitable energy management strategy for a hybrid energy system in a remote rural area of India. Energy, 94, 352–366.

[8] Rodríguez, J., Lai, J.-S., & Peng, F. Z. (2002). Multilevel inverters: A survey of topologies, controls, and applications. IEEE Transactions on Industrial Electronics, 49(4), 724–738.

[9] Srivastava, A. C., & Tripathi, A. (2024). Design and performance evaluation of a novel modular asymmetrical multilevel inverter with minimal switches, e-Prime. Advances in Electrical Engineering, Electronics and Energy.

[10] Holmes, D. G., & Lipo, T. A. (2003). *Pulse width modulation for power converters: Principles and practice*. IEEE Press.

[11] Singh, V., & Singh, R. S. (2018). A comparative study of multilevel inverter topologies, International Research Journal of Engineering Technology (IRJET), 5.

[12] Sadanala, C., Pattnaik, S., & Singh, V. P. (2022). A flying capacitor-based multilevel inverter architecture with symmetrical and asymmetrical configurations.

The IEEE Journal of Emerging and Selected Topics in Power Electronics, 10(2), 2210–2222.

[13] MathWorks. (2022). *FFT Analysis Using MATLAB*. MathWorks Documentation. https://www.mathworks.com/help/matlab/ref/fft.html.

[14] Bhandari, P., Padalia, D., Ranakoti, L., Khargotra, R., András, K., & Singh, T. (2023). Thermo-hydraulic investigation of open micro prism pin fin heat sink having varying prism sides. Alexandria Engineering Journal, 69, 457–468. https://doi.org/10.1016/j.aej.2023.02.016.

[15] Rekioua, D., Mokrani, Z., Kakouche, K., Rekioua, T., Oubelaid, A., Logerais, P. O., & Ghoneim, S. S. (2023). Optimization and intelligent power management control for an autonomous hybrid wind turbine photovoltaic diesel generator with batteries. Scientific Reports, 13(1), 21830. http://dx.doi.org/10.1038/s41598-023-49067-4.

[16] Sathish, T., Saravanan, R., Jeyanthi, C. E., Sabariraj, R. V., Hemachandu, P., Ağbulut, Ü., Saleel, A., & Soudagar, M. E. M. (2023). PCM embedded square channel receiver for effective operation of Net Zero Emission/Energy PDC building heating system. Sustainable Energy Technologies and Assessments, 60, 103519. http://dx.doi.org/10.1016/j.seta.2023.103519.

[17] Kanouni, B., Badoud, A. E., Mekhilef, S., Bajaj, M., & Zaitsev, I. (2024). Advanced efficient energy management strategy based on state machine control for multi-sources PV-PEMFC-batteries system. Scientific Reports, 14(1), 7996. https://doi.org/10.1038/s41598-024-58785-2.

[18] Almutairi, A., Albagami, N., Almesned, S., Alrumayh, O., & Malik, H. (2023). Electric vehicle load estimation at home and workplace in Saudi Arabia for grid planners and policy makers. Sustainability, 15(22), 15878. http://dx.doi.org/10.3390/su152215878.

[19] Yadav, D., Awasthi, M. K., & Kumar, A. (2025). *Next Generation Renewable Thermal Energy Harvesting, Conversion & Storage Technologies*. Elsevier.

[20] Kumar, A., Dutt, N., & Awasthi, M. K. (Eds.). (2024). *Heat Transfer Enhancement Techniques: Thermal Performance, Optimization and Applications*. John Wiley & Sons, Inc. https://doi.org/10.1002/9781394270996.

Chapter 9

Environmental Performance Evaluation of an Agrivoltaic System in a Hot Arid Region

Digvijay Singh, A.K. Singh, and Surrendra Poonia

9.1 INTRODUCTION

With the ever-growing global population, there is huge pressure on energy and food. The need for energy, water, and food combined can pose a big challenge. Agriculture alone is vulnerable to climate change as spatial and temporal variability, along with the absence of rainfall, hampers productivity. Therefore, environment-friendly, energy-efficient ways are required to ensure judicious use of energy in food production.

Agrivoltaic systems ensure the combination of growing crops in underutilized spaces in addition to generating electricity (Dinesh & Pearce, 2016; Mamun et al., 2022). The conventional electricity generation from the burning of coal results in considerably high amounts of GHG emissions.

In view of the aforementioned challenges, agrivoltaics is an option to protect the environment. India is the world's third-largest GHG producer/emitter, following the United States and China, producing 1.8 Gt CO_2 during 2021 (Dechamps, 2023) through coal burning. It has set a target to achieve NZE (Net Zero Emissions) by 2030, surpassing coal-fired power by the year 2027 (Obiora et al., 2024).

Agriculture itself is the main GHG producer. The PV electricity generation system is preferred to agriculture because its efficiency (15%) is much higher than that of crop photosynthesis (3%). Therefore, PV installation in agricultural land requires crop cultivation in interspaces and below panel areas (Poonia et al., 2022; Kumar et al., 2024b).

Additionally, agrivoltaic systems have numerous advantages in addition to power production and land utilization efficiency. Incorporating solar panels with farming, agrivoltaic systems may achieve partial shading, thus relieving heat stress for plants and limiting water evaporation, which is highly useful in hot and dry conditions of arid and semi-arid environments (Shahare et al., 2023; Datta et al., 2023; Wongchai et al., 2022; Rizwanuddin et al., 2023). Further, microclimatic alteration under photovoltaic arrays may result in higher retention of soil moisture as well as plant growth optimization, hence enhanced output for specific crops. Consequently, agrivoltaics

DOI: 10.1201/9781003634737-9

offers a potential solution to tackle the food-energy-water nexus and, at the same time, support climate change mitigation and sustainable agriculture (Raihan et al., 2024).

Keeping these points in mind, various designs of APV have been suggested. The APV system can probably have all the resources, namely energy, food, water, and CO_2 emissions (Anjum et al., 2022; Sharma et al., 2021; Rawat et al., 2022; Sathish et al., 2023; Bhandari et al., 2023; Uniyal et al., 2021).

9.2 METHODOLOGY

9.2.1 Objective

The main objective of this study is to compare this renewable form of energy source (PV system) with the coal-based thermal power source. Various designs, viz., G/M (Original), agrivoltaic designs, were considered for the prominent production of energy and agricultural produce (Mohanty et al., 2023; Kaswan et al., 2022; Pansare et al., 2023; Kanwar et al., 2023).

Several designs of the agrivoltaics system were analyzed for generating electricity and producing an ample harvest of crops (Poonia et al., 2022).

1. Design I: Traditional Ground-mounted PV (Tr-PV)
2. Design II: Traditional Ground-mounted PV with Agriculture. (Tr-PV + Ag)
3. Design III: One row full density with Agriculture ORFD + Ag
4. Design IV: Two row full density plus half density with Agriculture (TR(FD+HD)
5. Design V: Agriculture (Moth bean + Chili) (Agriculture)

In this study, a polycrystalline PV system with an efficiency of 15% and a sharp lifetime of 25 years was considered for comparing energy production efficiency (kWh/ha) and GWP potential (kg CO_2/ha).

9.2.2 System Design

Table 9.1 depicts Design I as a conventional setup that focuses on maximizing solar energy generation, while Design II integrates agrivoltaics to support crop growth in interspace of the PV panels. Design III increases the interspacing to 3.2m to optimize light distribution for both crops and PV efficiency. Design IV further expands the spacing to 6.4m to balance energy production with various farming needs. Design V serves as a control, consisting of a 1-hectare agricultural field without PV installations for baseline comparison (Thapliyal et al., 2023; Kumar et al., 2024a, 2024b; Singla et al., 2025).

Table 9.1 Agrivoltaic System Design Configurations, Including Interspacing and Design Purpose

Design	*Description*	*Interspacing*	*Purpose*
Tr-PV Design I	Traditional PV system	2m	Maximizing solar energy generation
Tr-PV + Ag Design II	Agrivoltaic system with crop cultivation	2m	Dual-use for energy and agriculture
ORFD + Ag Design III	Optimized interspacing for light distribution	3.2m	Enhancing crop growth & PV efficiency
TR(FD+HD) Design IV	Wider interspacing for better farming flexibility	6.4m	Balancing farming needs & PV production
Agriculture Design V	Control setup (no PV)	N/A	Baseline agricultural comparison

Table 9.2 Energy Equivalents of Various Agricultural Inputs Providing Quantitative Representation of Different Inputs

Input	*Unit*	*Energy Equivalent (kWh)*
Man	man-hours	0.544
Diesel	lit	15.34
Self-Propelled Machine	kg	18
Farm Machinery	kg	17.42
Fertilizer, Nitrogen	kg	16.7
P_2O_5	–	3.1
K_2O	–	1.86
Pulse	–	4.08
Chili (Crop/Kg)	kg	1.56

9.2.3 Crop Selection

The crops have been selected due to their adaptability in arid and semi-arid regions. Mothbean legume can withstand drought conditions with little irrigation, whereas the cash crop chili has a strong market demand (Patidar & Singh, 2000).

The energy input and output of crops of MG and chili were computed using the energy equivalents of various resources such as manpower (human), diesel, electricity, seeds, FYM, fertilizers, and machinery, both source-wise and operation-wise (Table 9.2). The output energy was determined as per the yield and energy ratio (output energy/input energy) (Singh et al., 2007).

Table 9.3 Embodied Energy of Various Components in PV System Listing Silicon Crystal Formation, Solar Cell Manufacturing, Lamination, Maintenance, Administration, Structural Support, Inverter, and Wiring.

S. No.	*Components*	*Embodied energy (kWh/kWᵖ)*
1	Silicon crystal formation (metallurgy & electronic-grade purification)	670
2	Manufacturing of solar cell	120
3	Lamination and assembly	190
4	Maintenance (labor)	984
5	Administration	53.5
6	Iron stand, screws, installation, & support system	53
7	Inverter/kW	210
8	Wires/m	3

9.2.4 Embodied Energy Calculation of PV Systems

The embodied energy of the PV system was calculated based on the energy required for the production of silicon crystal and manufacturing of solar cell, laminate, glass, labor, aluminum installation, inverters, wires, etc. (Table 9.3). This was spread over the life of the PV cell (25 years), and annual embodied energy was considered (Rajput & Singh, 2017; Singh et al., 2022).

9.2.5 Global Warming Potential (GWP) and Energy Savings Calculation

Based on the embodied energy of PV production and input energy of crops, the global warming potential of different designs was calculated (Kg CO_2 e/kWh) using Equations 9.1 and 9.2. Based on the monitoring of the GMPV system installed at CAZRI Jodhpur, it was observed that one kWp PV installation (375 kW and 105 kW) can produce about 4 kWh per day annually. EPBP for PV and crops calculated using Equations 9.3 and 9.4. The electricity production of different APV designs has been estimated accordingly (Poonia et al., 2022). Total annual energy saving was calculated using the quantum of PV electricity generation and crop output energy and subtracting the embodied energy of crop and input energy (see Equation 9.5).

$$\left(\frac{GWP}{kWh}\right)_{PV} = \frac{Annual\,embodied\,energy}{Annual\,PV\,electricity\,generation} \quad (9.1)$$

$$\left(\frac{GWP}{kWh}\right)_{Crop} = \frac{Input\,energy\,of\,crop}{Output\,energy\,of\,crop} \quad (9.2)$$

$$EPBP_{PV} = \frac{Total\,embodied\,energy}{Annual\,PV\,production} \quad (9.3)$$

$$EPBP_{Crop} = \frac{Total\,input\,energy}{output\,energy} \quad (9.4)$$

$$Energy\,Saving = Total\,Energy\,Output - Total\,Energy\,input \quad (9.5)$$

9.3 RESULT

It is evident from Table 9.4 that GWP (Kg CO_2e/ha) and total energy production (kWh/ha/yr) values suggest that the best design is Design II for total energy production with different values of GWP (209071 for MB and 213139 for chili). Table 9.5 provides the details of energy input and output

Table 9.4 Annual Global Warming Potential (GWP) (Kg CO_2 E/Ha/Yr) and Total Energy Production (Kwh/Ha/Yr) for Different Design Variants and Crops

Production System	*GWP (Kg CO_2e/ha/yr)*	*Total Energy Production (kWh/ha/yr)*
Crop (Moth Bean)	**1,921**	**6,726**
Design I	208,263	1,200,000
Design II	209,071	1,202,290
Design III	132,389	758,621
Design IV	104,773	598,763
Crop (Chili)	**8,682**	**17,364**
Design I	208,263	1,200,000
Design II	213,136	1,209,725
Design III	137,460	768,527
Design IV	109,963	609,120

Table 9.5 Energy Input and Output of Moth Bean and Chili (Kwh/Ha) and Their Combination of Crops

Crop	*Mothbean*	*Chili*	*MB + Chili*
Operation wise	790.8	3,805	4,595.8
Source wise	1,131	4,877	6,008
Total input energy	1921.8	8,682	10,603
Total output energy	6,726	17,364	24,090
Energy gain (kWh/yr)	4,804	8,682	13,486
GWP/kWh (kg CO_2/kWh)	0.286	0.50	0.44

Table 9.6 Energy Pay Back Period (EPBP) & GWP/kWh Generated for Different Designs Compared to Coal Power

	Mothbean		*Chili*	
Production System	*GWP (Kg CO_2 e)*	*EPBP (yr)*	*GWP/kWh*	*EPBP (yr)*
Design I	0.17	4.34	0.175	4.34
Design II	0.17	4.35	0.176	4.4
Design III	0.175	4.36	0.179	4.47
Design V	0.32	0.31	0.46	0.46
Design IV	0.175	4.38	0.181	4.51
Coal Power	**1.0**	–	**1.0**	–

Table 9.7 Annual Energy Savings (kWh/ha) of different Production Systems, Including Moth Bean, Chili, and Combination

Production System	*Without Crops*	*Moth Bean*	*Chili*	*Moth Bean + Chili*
Design I	991,737	N/A	N/A	N/A
Design II	N/A	993,220	996,595	998,076
Design III		626,232	631,111	634,000
Design V		4,805	8,682	13,468
Design IV		493,970	499,157	502,307

for agriculture crops (mothbean & chili) with lower energy gain and higher GWP/kWh as compared to traditional PV.

Table 9.6 depicts EPBP of all designs varied from 4.34 to 4.51 years, whereas GWP was minimum for Trop. PV (0.17 kg CO_2 e/kWh) and maximum for Design III (0.179) with chili crop. Design V had slightly GWP (0.32) for MB and 0.46 for chili with minimum energy EPBP of 0.318 and 0.46 years for MB and chili, respectively, which shows that only agriculture has high value of GWP. This is the reason why agrivoltaic systems have slightly higher GWP than those of Trop. PV. All the designs are better than coal power input, which has GWP (Kg CO_2 e/kWh) as high as 1.0, indicating that for generating 1 kWh, the coal power unit will emit 1.0 kg of CO_2.

However, Table 9.7 provides values of energy saving (kWh/ha) for different production systems. The best option was Design II (Moth Bean + Chili) with the highest energy saving (998076 kWh/ha). Table 9.8 provides the details of APV designs with interspaces, capacity, annual PV generation, and total generation for all the designs. The highest for Design II (MB + chili) was 1212015 kWh/yr.

Table 9.8 Comparison of Different Agrivoltaic Design Configurations Based on Inter-Space Distance, Photovoltaic (PV) System Capacity Per Hectare, and Annual Energy Generation from PV and Crop Production

Design	*Design I*	*Design II*	*Design III*	*Design IV*	*Design V*
Inter space (m)	2.0	2.0	3.2	6.4	–
PV system (kW) capacity/ha	765	765	482	380	–
Annual PV generation (kWh)	1,200,000	1,200,000	755,555	595,620	6,726 (MB) 17,364 (Chili)
Total generation (kWh/ha)	1,200,000	120,2290 (MB)	758,621	598,763	
		120,9725(Chili)	768,577	609,120	
MB (mothbean)					
MB (mothbean) + chili	1,200,000	1,212,015	771,643	612,260	24,090

9.4 CONCLUSIONS AND FUTURE SCOPE

The APV system is a viable option for the simultaneous production of crops and electricity. This system can make the best use of fertile agricultural land in general and hot arid zones (average annual daily irradiance of 6 kWh/m^2) in particular. Several APV designs with different inter-row spaces were evaluated for their performance. The Design II (Tr-PV + Ag) was found to be the best option, followed by Design III (ORFD + Ag).

In addition to crop and electricity production, the APV system can also be used for harvesting rainwater, which is collected in a water tank for storage. This water can be used for cleaning solar panels in addition to providing drip irrigation to crops. The shading effect of panels can also reduce the evaporation and transpiration requirements of crops. Shade-loving crops could also be grown in the below-panel area.

This APV system can also be optimized for various climatic zones of the country. Based on the performance of designs, the best design can be recommended. This APV system can play a big role in mitigating GHG emissions and can provide an eco-friendly environment by reducing pollution.

REFERENCES

Al Mamun, M. A., Dargusch, P., Wadley, D., Zulkarnain, N. A., & Aziz, A. A. (2022). A review of research on agrivoltaic systems. *Renewable and Sustainable Energy Reviews*, 161, 112351.

Anjum, S., Rana, S., Dasila, K., Agnihotri, V., Pandey, A., & Pande, V. (2022). Comparative nutritional and antimicrobial analysis of Himalayan black and yellow soybean and their okara. *Journal of the Science of Food and Agriculture*, 102(12), 5358–5367. http://dx.doi.org/10.1002/jsfa.11889.

Bhandari, P., Padalia, D., Ranakoti, L., Khargotra, R., András, K., & Singh, T. (2023). Thermo-hydraulic investigation of open micro prism pin fin heat sink having varying prism sides. *Alexandria Engineering Journal*, 69, 457–468. https://doi.org/10.1016/j.aej.2023.02.016.

Datta, S., Hamim, I., Jaiswal, D. K., et al. (2023). Sustainable agriculture. *BMC Plant Biology*, 23, 588. https://doi.org/10.1186/s12870-023-04626-9.

Dechamps, P. (2023). The IEA World Energy Outlook 2022–a brief analysis and implications. *European Energy & Climate Journal*, 11(3), 100–103.

Dinesh, H., & Pearce, J. M. (2016). The potential of agrivoltaic systems. *Renewable and Sustainable Energy Reviews*, 54, 299–308.

Kanwar, N., Jadoun, V. K., Jayalakshmi, N. S., Afthanorhan, A., Fatema, N., Malik, H., & Hossaini, M. A. (2023). Industry—challenge to pro-environmental manufacturing of goods replacing single-use plastic by Indian industry: A study toward failing ban on single-use plastic access. *IEEE Access*, 11, 77336–77346. http://dx.doi.org/10.1109/ACCESS.2023.3296097.

Kaswan, M. S., Rathi, R., Cross, J., Garza-Reyes, J. A., Antony, J., & Yadav, V. (2022). Integrating green lean six sigma and industry 4.0: a conceptual framework. *Journal of Manufacturing Technology Management*, 34(1), 87–121. http://dx.doi.org/10.1108/JMTM-03-2022-0115.

Kumar, A., Awasthi, K. M., & Dutt, N. (2024a). Nanofluidic applications in sustainable food technology and agriculture. In: Awasthi, M. K. (ed.) *Nanofluid Dynamics and Transport Phenomenon*. Boca Raton, FL: CRC Press, pp. 224–242. https://doi.org/10.1201/9781003527251-13.

Kumar, A., Singla, Y., & Namboodri, T. (2024b). Globalization and international issues in sustainable manufacturing. In: Shah, S., Nautiyal, H., Gugliani, G., Kumar, A., Namboodri, T., & Y. K. Singla (eds.) *Sustainability in Smart Manufacturing: Trends, Scope and Challenges*. Boca Raton, FL: CRC Press, pp. 1–18. https://doi.org/10.1201/9781003467496-1.

Mohanty, S., Panwar, V., Anoop, G., & Park, S. (2023). Eco-friendly phyllanthus emblica-based ionic polymer composite for enhanced mechanical, electrical, and wearable sensing performance. *Materials Chemistry and Physics*, 303, 127791. https://doi.org/10.1016/j.matchemphys.2023.127791.

Obiora, S. C., Bamisile, O., Hu, Y., Ozsahin, D. U., & Adun, H. (2024). Assessing the decarbonization of electricity generation in major emitting countries by 2030 and 2050: Transition to a high share renewable energy mix. *Heliyon*, 10(8).

Pansare, R., Yadav, G., Garza-Reyes, J. A., & Raosaheb Nagare, M. (2023). Assessment of sustainable development goals through Industry 4.0 and reconfigurable manufacturing system practices. *Journal of Manufacturing Technology Management*, 34(3), 383–413. http://dx.doi.org/10.1108/JMTM-05-2022-0206.

Patidar, M., & Singh, A. K. (2000). Effect of tillage practices on yield and energy use pattern of clusterbean (Cyamopsis tetragonoloba L. Taub)-pearl millet (Pennisetum glaucum LR Br.) rotation in arid zone. *Journal of Agricultural Engineering*, 37(4), 1–6.

Poonia, S., Jat, N. K., Santra, P., Singh, A. K., Jain, D., & Meena, H. M. (2022). Techno-economic evaluation of different agri-voltaic designs for the hot arid ecosystem India. *Renewable Energy*, 184, 149–163.

Raihan, A., Zimon, G., Alam, M. M., Khan, M. R., & Sadowska, B. (2024). Nexus between nuclear energy, economic growth, and greenhouse gas emissions in India. *International Journal of Energy Economics and Policy*, 14(2), 172–182.

Rajput, S. K., & Singh, O. (2017, November). Reduction in CO2 emission through photovoltaic system: a case study. In *Proceedings of the 3rd IEEE International Conference on Nanotechnology for Instrumentation and Measurement*, Greater Noida, India (pp. 16–17).

Rawat, B., Rawat, J. M., Purohit, S., Singh, G., Sharma, P. K., Chandra, A., . . . Qureshi, K. A. (2022). A comprehensive review of Quercus semecarpifolia Sm.: An ecologically and commercially important Himalayan tree. *Frontiers in Ecology and Evolution*, 10, 961345–961351. http://dx.doi.org/10.3389/fevo.2022.961345.

Rizwanuddin, S., Kumar, V., Naik, B., Singh, P., Mishra, S., Rustagi, S., & Kumar, V. (2023). Microbial phytase: Their sources, production, and role in the enhancement of nutritional aspects of food and feed additives. *Journal of Agriculture and Food Research*, 12, 100559. https://doi.org/10.1016/j.jafr.2023.100559.

Sathish, T., Saravanan, R., Jeyanthi, C. E., Sabariraj, R. V., Hemachandu, P., Ağbulut, Ü., . . . Soudagar, M. E. M. (2023). PCM embedded square channel receiver for effective operation of net zero emission/energy PDC building heating system. *Sustainable Energy Technologies and Assessments*, 60, 103519. http://dx.doi.org/10.1016/j.seta.2023.103519.

Shahare, Y., Singh, M. P., Singh, P., Diwakar, M., Singh, V., Kadry, S., & Sevcik, L. (2023). A comprehensive analysis of machine learning-based assessment and prediction of soil enzyme activity. *Agriculture*, 13(7), 1323. https://doi.org/10.3390/agriculture13071323.

Sharma, P. K., Rausa, K., Rani, A., Mukherjee, S., & Kumar, M. (2021). Biopurification of dairy farm wastewater through hybrid constructed wetland system: Groundwater quality and health implications. *Environmental Research*, 200, 111426. http://dx.doi.org/10.1016/j.envres.2021.111426.

Singh, D., Poonia, S., & Singh, A. K. (2022). Carbon reduction and economic evaluation of building attached photovoltaic systems. *Materials Today: Proceedings*, 63, 92–98.

Singh, H., Singh, A. K., Kushwaha, H. L., & Singh, A. (2007). Energy consumption pattern of wheat production in India. *Energy*, 32(10), 1848–1854.

Singla, Y. K., Kumar, A., & Maughan, M. (2025). *Digital manufacturing in Industry 5.0*. Publisher Springer Nature.

Thapliyal, S., Wazid, M., Singh, D. P., Das, A. K., Shetty, S., & Alqahtani, A. (2023). Design of robust Blockchain-envisioned authenticated key management mechanism for smart healthcare applications. *IEEE Access*. https://doi.org/10.1109/ACCESS.2023.3310264.

Uniyal, S., Mangla, S. K., Sarma, P. R., Tseng, M. L., & Patil, P. (2021). ICT as "Knowledge management" for assessing sustainable consumption and production in supply chains. *Journal of Global Information Management (JGIM)*, 29(1), 164–198. http://dx.doi.org/10.4018/JGIM.2021010109.

Wongchai, A., Shukla, S. K., Ahmed, M. A., Sakthi, U., & Jagdish, M. (2022). Artificial intelligence-enabled soft sensor and internet of things for sustainable agriculture using ensemble deep learning architecture. *Computers and Electrical Engineering*, 102, 108128. https://doi.org/10.1016/j.compeleceng.2022.108128.

Chapter 10

Next-Generation Solar Power Plants

Technological Developments, Challenges, and Future Aspects

Neha Singh Raghuvanshi, Ashwani Kumar, Bharti Sharma, Yatika Gori, Sakshi Sarathe, and Gaurav Dwivedi

10.1 INTRODUCTION

The growing global energy demand and the escalating challenges of climate change have underscored the urgency of transitioning to renewable energy sources. Among these, solar energy has emerged as a cornerstone of sustainable development due to its abundance, accessibility, and environmental benefits. The sun, as an inexhaustible source of energy, radiates more power onto the Earth's surface in an hour than humanity consumes in an entire year. Harnessing this energy effectively is pivotal in reducing our dependence on fossil fuels, which are not only finite but also significant contributors to greenhouse gas emissions and environmental degradation. Solar power stands out as a clean and renewable energy source capable of addressing the dual objectives of meeting energy demands and mitigating climate change. Unlike fossil fuels, which release carbon dioxide and other pollutants during combustion, solar energy systems generate electricity with minimal environmental impact. This advantage is particularly critical in the context of international efforts to limit global temperature rise and achieve net-zero emissions. The deployment of solar technologies is a testament to humanity's ability to innovate and adapt to the pressing need for sustainable energy solutions. The journey of solar power from its inception to its current state of advancement has been remarkable. Early solar power systems relied on simple photovoltaic cells, primarily used in niche applications such as powering calculators or satellites. These systems, while ground-breaking, were limited by low efficiency and high production costs, making large-scale adoption economically unviable. However, persistent research and technological breakthroughs over the decades have revolutionized the field. Modern photovoltaic systems are not only far more efficient but also significantly more affordable, making them a practical choice for widespread use in both urban and rural areas [1].

 DOI: 10.1201/9781003634737-10

Concentrated solar power (CSP) systems have also played a transformative role in the evolution of solar energy. These systems, which use mirrors or lenses to focus sunlight onto a small area to produce heat, offer an alternative approach to electricity generation. CSP technology has seen substantial improvements in recent years, with innovations such as advanced thermal storage systems enabling the generation of electricity even during periods of low sunlight or at night. This adaptability has increased the reliability and appeal of solar power as a primary energy source. The shift from traditional solar power plants to next-generation installations marks a new era in the renewable energy landscape [2]. Today's solar power plants are characterized by their integration with cutting-edge technologies, including artificial intelligence, machine learning, and energy storage solutions. These advancements have enhanced the efficiency, scalability, and functionality of solar systems, allowing them to meet the dynamic demands of modern energy grids. Innovations such as floating solar farms, agrivoltaics, and building-integrated photovoltaics further illustrate the versatility and potential of solar energy in diverse contexts. The adoption of solar power on a global scale is not merely a technological achievement but also a profound step toward a sustainable future. By leveraging the power of the sun, humanity can reduce its carbon footprint, enhance rudimentary beginnings to the sophisticated systems of today, and reflect the relentless pursuit of a cleaner and more equitable energy paradigm. As the world continues to grapple with the challenges of climate change, the role of solar energy as a cornerstone of sustainable development becomes increasingly indispensable.

10.2 TECHNOLOGICAL ADVANCEMENTS IN SOLAR POWER PLANTS

Photovoltaic (PV) technologies are at the heart of solar power generation, converting sunlight directly into electricity using semiconductors. Over the decades, these technologies have undergone remarkable advancements, improving their efficiency, cost-effectiveness, and versatility. Next, we explore the key PV technologies, their features, advantages, and limitations, supplemented with examples and a comparative table.

10.2.1 Photovoltaic (PV) Technologies

10.2.1.1 Mono-Si, Poly-Si, and Thin-Film Technologies

10.2.1.1.1 Monocrystalline Silicon (Mono-Si)

Monocrystalline silicon solar cells are made from a single, continuous crystal structure, resulting in high efficiency and durability. They typically exhibit efficiencies ranging from 20% to 24% under standard conditions, making them one of the most efficient solar cell types available. Their distinct black appearance, resulting from the uniform silicon composition, enhances their aesthetic appeal for rooftop installations. SunPower's Maxeon solar panels, which leverage mono-Si technology, are known for their high efficiency and long warranties [3].

Modern solar technologies, including monocrystalline and tandem solar cells, deliver high energy conversion rates, maximizing output while minimizing the physical footprint. This makes them ideal for urban installations with limited space. With a durability range of 25–30 years, solar panels offer reliable energy generation, providing consistent returns on investment and reducing the need for frequent replacements. Advanced solar cells are optimized to generate electricity even under cloudy skies or shaded areas, ensuring dependable energy supply across varying weather conditions. Advanced solar technologies, like monocrystalline and tandem cells, require complex manufacturing processes and premium materials, leading to elevated production costs compared to conventional energy sources. The manufacturing process often results in significant material wastage, especially in cutting silicon wafers or producing thin-film panels, impacting overall efficiency and sustainability (Table 10.1).

10.2.1.1.2 Polycrystalline Silicon (Poly-Si)

Poly-Si solar cells are made from multiple silicon crystals, making them less expensive to produce than mono-Si cells. They are recognizable by their blue, speckled appearance and generally have efficiencies ranging from 15% to 19%. While slightly less efficient than mono-Si, they remain a popular choice for large-scale installations due to their cost-effectiveness. Canadian Solar's polycrystalline panels are widely used in utility-scale projects [4].

Table 10.1 Comparative Analysis of Mono-Si, Poly-Si, and Thin-film Technologies

Technology	*Efficiency %*	*Lifespan (yrs)*	*Cost*	*Applications*
Monocrystalline (Mono-Si)	20–24	25–30	High	Rooftop and premium projects
Polycrystalline (Poly-Si)	15–19	20–25	Moderate	Utility-scale installations
Thin-film	10–16	10–20	Low	Curved surfaces, portable systems

Technologies like polycrystalline and thin-film solar panels are more affordable to produce, making them accessible for large-scale installations and residential use. While not as efficient as premium technologies, these systems strike a balance between cost and performance, offering a reliable and economical option for diverse energy needs. Polycrystalline and thin-film panels require more space to generate the same power output as monocrystalline panels, making them less ideal for areas with limited space. These panels typically have a slightly reduced lifespan compared to monocrystalline panels, requiring replacements or maintenance sooner, which may affect long-term cost efficiency (Table 10.1).

10.2.1.1.3 Thin-Film Technologies

Thin-film solar cells are made by depositing one or more layers of photovoltaic material onto a substrate such as glass, plastic, or metal. Common materials include amorphous silicon, cadmium telluride (CdTe), and copper indium gallium selenide (CIGS). These panels are lightweight, flexible, and have efficiencies typically ranging from 10% to 16%. First Solar is a leading manufacturer of CdTe thin-film panels used in utility-scale projects [5].

Thin-film solar panels are adaptable to non-traditional applications, including curved surfaces, rooftops with structural limitations, and portable solar solutions. Their production requires less material and energy, making them more cost-effective than crystalline silicon panels, ideal for large-scale and budget-conscious projects. Thin-film panels typically have lower energy conversion rates, requiring more area to produce the same power output as crystalline counterparts. Certain thin-film technologies, like cadmium telluride (CdTe), utilize toxic materials, increasing recycling and disposal challenges and necessitating stricter environmental safeguards (Table 10.1).

10.2.1.1.3.1 PEROVSKITE SOLAR CELLS

Perovskite solar cells represent one of the most promising advancements in PV technology, with efficiencies surpassing 25% in laboratory settings. These cells use a perovskite-structured compound (e.g., methylammonium lead halides) as the light-absorbing layer. The material's excellent light-harvesting capabilities and low manufacturing costs make it an exciting candidate for next-generation solar cells. Oxford PV is a leader in commercializing perovskite-on-silicon tandem solar cells, showcasing their potential for commercial applications [6].

Rapid progress has led to lab-record efficiencies surpassing traditional silicon cells. These can be manufactured using low-temperature processes and printed onto various substrates, offering compatibility with lightweight, flexible materials for innovative applications like wearable solar cells or building-integrated photovoltaics (BIPVs). Perovskite cells are prone to degradation under moisture, heat, and prolonged UV exposure, limiting their practical lifespan. Many perovskite materials contain lead, raising environmental and health concerns. Challenges remain in scaling laboratory achievements to mass production while maintaining performance.

10.2.1.1.3.2 TANDEM SOLAR CELLS FOR HIGHER EFFICIENCY

Tandem solar cells combine multiple layers of photovoltaic materials, each optimized to absorb different parts of the solar spectrum. By utilizing the strengths of various materials, these cells achieve efficiencies significantly higher than single-junction cells. For instance, tandem cells that pair silicon with perovskite have demonstrated efficiencies exceeding 30% in laboratory tests. Example: The National Renewable Energy Laboratory (NREL) developed a record-breaking tandem solar cell with a 39.5% efficiency under one-sun illumination in 2022 [7].

By addressing the limitations of single-junction cells, tandem designs achieve better energy conversion. Each layer in a tandem cell targets specific wavelengths of light for maximum energy capture. Despite higher upfront costs, tandem cells offer long-term savings due to higher energy output. Tandem cells require precise layering and alignment of materials, increasing production complexity. Ensuring stability and performance when combining different materials is a significant hurdle (Table 10.2).

Photovoltaic technologies have diversified significantly to meet the evolving demands of solar power generation. From traditional silicon-based cells to cutting-edge tandem and perovskite designs, the industry is witnessing a revolution in efficiency, affordability, and versatility. As these advancements transition from laboratories to large-scale deployments, they hold immense potential to drive the global shift toward sustainable energy solutions.

Table 10.2 Case Comparison of Single-Junction versus Tandem Solar Cells

Type	*Efficiency %*	*Advantages*	*Limitations*
Single Junctions	15–25	Simpler manufacturing, lower cost	Limited by Shockley-Queisser limit
Tandem	30–40	High efficiency, optimal spectrum use	Complex production, higher cost

Table 10.3 CSP Technologies: Efficiency, Advantage, and Applications

Technology	*Efficiency %*	*Advantages*	*Applications*
Parabolic Troughs	**15–20**	Proven, scalable, and reliable storage	Large-scale solar plants
Solar Towers	**20–30**	High efficiency, compact, superior storage	High-temperature industrial processes
Fresnel Reflectors	**8–12**	Cost-effective, simpler design	Small- to medium-scale installations

10.2.2 Concentrated Solar Power (CSP) Innovations

Concentrated solar power (CSP) technologies focus sunlight onto a small area to generate high temperatures, which drive a heat engine or a turbine connected to an electrical generator. Unlike photovoltaic (PV) systems, which directly convert sunlight into electricity, CSP systems harness solar thermal energy, making them highly efficient for large-scale energy production and thermal storage. CSP innovations have enhanced efficiency, scalability, and adaptability, positioning these systems as critical players in the transition to renewable energy [8] (Table 10.3).

10.2.2.1 Parabolic Troughs, Solar Towers, and Fresnel Reflectors

10.2.2.1.1 Parabolic Troughs

Parabolic trough systems are the most widely deployed CSP technology. They use curved mirrors to focus sunlight onto a receiver tube containing a heat-transfer fluid, typically synthetic oil or molten salt. The fluid transfers the heat to a steam turbine for electricity generation. The Andasol Solar Power Station in Spain, one of the largest parabolic trough systems, generates 150 MW of power, supplying electricity to approximately 200,000 households [9]. CSP systems, such as parabolic troughs and solar towers, have a long history of reliable performance, with well-established operational frameworks. CSP effectively stores energy using molten salts, enabling power generation during non-sunny hours and providing grid stability. CSP plants can be expanded in stages, making them suitable for both medium- and large-scale energy needs. CSP installations need extensive land with direct sunlight, limiting deployment in densely populated or geographically constrained areas. Unlike PV systems, CSP relies on direct sunlight, making performance suboptimal in regions with frequent cloud cover or diffuse sunlight (Table 10.3).

10.2.2.1.2 Solar Towers

Solar tower systems (or central receiver systems) use a field of heliostats (mirrors) to direct sunlight onto a central receiver located on a tower. The receiver absorbs the concentrated sunlight and heats a working fluid, typically molten salt, which then generates steam for power production. The Ivanpah Solar Electric Generating System in California produces 392 MW of electricity, enough to power 140,000 homes, using solar tower technology [10]. Solar towers achieve higher efficiency by operating at elevated temperatures, enhancing power conversion and overall system performance. Compared to parabolic troughs, solar towers require a smaller footprint, making them more suitable for concentrated installations. Their integration with molten salt storage systems ensures effective energy storage, allowing continuous power generation during non-sunny periods. The precise alignment and management of heliostats (mirrors directing sunlight to the tower) require advanced technology, increasing system complexity and maintenance demands. Solar towers entail significant upfront investment, making them costlier than other CSP technologies and potentially delaying the return on investment (Table 10.3).

10.2.2.1.3 Fresnel Reflectors

Fresnel reflector systems use flat or slightly curved mirrors to focus sunlight onto a linear receiver. They offer a simpler and more cost-effective alternative to parabolic troughs while maintaining similar functionality. Example: The Liddell Power Station in Australia incorporates a Fresnel CSP system to augment its coal-fired power generation, showcasing hybrid potential [11]. Fresnel reflectors have a simpler design and require less material, which reduces both production and maintenance costs compared to more complex CSP systems. The system's design is more compact and straightforward, making it easier to transport, install, and maintain, especially for small to medium-scale projects. While effective, Fresnel reflectors are less thermally efficient than parabolic troughs and solar towers, leading to lower overall energy output per unit area. To maintain optimal sunlight focus, Fresnel reflectors rely on precise tracking mechanisms, adding complexity and maintenance requirements to the system (Table 10.3).

10.2.3 Hybrid CSP-PV Systems

Hybrid systems combine concentrated solar power (CSP) with photovoltaic (PV) technologies to maximize the advantages of both approaches. These systems use PV panels for immediate electricity generation and CSP systems with thermal storage to provide power during periods of low sunlight or at night. By integrating the strengths of both technologies, hybrid CSP-PV systems achieve higher efficiency, reliability, and scalability. The Noor Midelt hybrid project in Morocco is a pioneering hybrid CSP-PV plant with a planned capacity of 800 MW. It combines PV panels for daytime generation

Table 10.4 Standalone CSP versus Hybrid CSP-PV

Parameter	*Standalone CSP*	*Hybrid-CSP-PV*
Energy Output	Limited to thermal output	Combined thermal and PV output
Cost	Higher LCOE	Optimized LCOE
Storage Capabilities	Superior	Superior with PV contribution
Operational Flexibility	Moderate	High

with CSP thermal storage for dispatchable energy, ensuring round-the-clock electricity supply [12]. PV panels produce electricity during the day, while CSP systems ensure power availability during off-peak sunlight hours. Combining CSP and PV can reduce the levelized cost of electricity (LCOE) by optimizing resource utilization. Hybrid systems address intermittency issues associated with standalone PV or CSP systems, improving grid integration (Table 10.4).

Innovations in CSP technologies, from parabolic troughs to hybrid CSP-PV systems, have expanded the potential of solar energy as a reliable and efficient source of power. Parabolic troughs and solar towers are well-suited for large-scale installations, while Fresnel reflectors offer cost-effective solutions for smaller projects. Hybrid CSP-PV systems exemplify the power of technological integration, delivering continuous, stable energy supply. As these innovations advance, CSP technologies will play a pivotal role in achieving a sustainable energy future.

10.2.4 Emerging Technologies

As the demand for more efficient and sustainable energy solutions grows, emerging technologies in solar energy are making significant strides in reshaping the future of power generation. These innovations promise to enhance the efficiency, cost-effectiveness, and flexibility of solar energy systems. Among the most promising of these technologies are quantum dot solar cells, organic photovoltaics (OPVs), and transparent solar panels. These advancements push the boundaries of traditional photovoltaic systems and offer new possibilities for integrating solar energy into various applications.

10.2.4.1 Quantum Dot Solar Cells

Quantum dots (QDs) are nanometer-sized semiconductor particles that possess unique optical and electronic properties due to quantum mechanics. When integrated into solar cells, quantum dots can enhance energy conversion efficiencies by exploiting multiple absorption spectra. Quantum dot solar cells (QDSCs) are gaining attention for their ability to improve

Table 10.5 Quantum Dot versus Traditional Solar Cells

Parameter	*Quantum Dot Solar Cells*	*Traditional Silicon Solar Cells*
Efficiency	Up to 30% (theoretical)	20–25% (commercial)
Cost	Lower potential due to cheaper materials	High (due to silicon wafer production)
Flexibility	Flexible, lightweight	Rigid, heavy
Durability	Limited by material degradation	High, 25+ years

performance while maintaining relatively low costs [13]. Quantum dots absorb light in a broad range of wavelengths, which allows them to harvest more energy from the sunlight spectrum. They can also be engineered to have specific bandgaps, enabling them to tune their absorption to different parts of the solar spectrum. This flexibility increases the overall efficiency of the solar cell. In 2020, researchers at the University of Chicago achieved a significant breakthrough in QDSCs by creating a solar cell that efficiently absorbs both visible and infrared light, potentially increasing the efficiency of quantum dot solar cells to 30%. Such advances are expected to revolutionize solar technology (Table 10.5).

10.2.4.2 Organic Photovoltaics

Organic photovoltaics (OPVs) are a type of solar cell that uses organic compounds (carbon-based materials) to convert sunlight into electricity. OPVs are particularly promising for applications that require lightweight, flexible, and cost-effective solar energy solutions. Their unique design offers potential for large-scale integration, especially in urban areas and portable devices [14]. In OPVs, sunlight excites electrons in the organic materials, creating electron-hole pairs that are separated and driven to the electrodes, generating an electric current. Unlike traditional silicon solar cells, OPVs are based on organic semiconductors such as polymers and small molecules. In 2021, researchers developed an OPV system with an efficiency of 18%, a significant improvement over earlier prototypes. This technology is being explored for applications like solar-powered wearables and portable electronics (Table 10.6).

OPVs can be manufactured on flexible substrates, making them ideal for integrating into various surfaces such as clothing, windows, and building materials. Organic materials are relatively inexpensive and can be processed using low-cost methods like printing. OPVs can be mass-produced using roll-to-roll printing techniques, making them scalable for various applications. OPVs typically offer lower efficiency (around 10–20%) compared to

Table 10.6 OPV versus Traditional PV Cells

Parameter	*Organic Photovoltaics*	*Traditional Silicon Solar Cells*
Efficiency	10–20%	20–25
Cost	Low (due to inexpensive materials)	High (due to silicon wafer production)
Flexibility	Highly Flexible, lightweight	Rigid, Heavy
Durability	Limited (due to organic degradation)	High, 25+ years

Table 10.7 Transparent Solar Panels versus Traditional Solar Panels

Parameter	*Transparent Solar Panels*	*Traditional Solar Panels*
Efficiency	6–8%	15–22%
Transparency	High (allows visible light)	Low (opaque)
Integration	Ideal for urban buildings	Suitable for rooftops and open areas
Cost	High (due to material complexity)	Moderate (due to mass production)

silicon-based cells. Organic materials degrade faster than silicon, limiting their operational lifespan (Table 10.6).

10.2.4.3 Transparent Solar Panels

Transparent solar panels are an emerging technology that allows sunlight to pass through while simultaneously generating electricity. These panels function as windows or glazing materials that generate solar power without obstructing views, making them highly promising for urban environments where space for traditional solar panels may be limited. Transparent solar panels use a special class of materials that absorb sunlight in the ultraviolet (UV) and infrared (IR) ranges, while allowing visible light to pass through. By selectively absorbing certain wavelengths of light, these panels can still generate electricity while maintaining transparency. In 2020, researchers at Michigan State University developed transparent solar panels that can achieve efficiencies of up to 8%, offering a potential solution for integrating solar energy into windows and building facades. These panels have the potential to turn all building windows into power-generating surfaces [15] (Table 10.7).

These panels can be integrated seamlessly into building designs without disrupting architectural aesthetics. They serve as both windows and energy producers, reducing the need for additional space. Ideal for cities with limited rooftop space, transparent solar panels can be used in building-integrated photovoltaics (BIPV) to maximize energy generation. Due to the selective absorption of light, these panels typically have lower efficiency compared to conventional solar panels. Manufacturing transparent solar panels can be expensive due to the advanced materials required for selective light absorption (Table 10.7).

The emergence of quantum dot solar cells, organic photovoltaics, and transparent solar panels represents a significant step forward in the evolution of solar energy technology. These innovations offer various advantages, including flexibility, cost-effectiveness, and integration potential but also come with challenges such as lower efficiency and material degradation. However, as research progresses, these technologies have the potential to revolutionize the solar energy landscape, creating new opportunities for solar power generation in urban environments, wearable devices, and beyond. Their development is crucial in the quest to make solar energy more accessible, versatile, and sustainable (Table 10.7).

10.3 ENERGY STORAGE SOLUTIONS

As solar energy continues to gain prominence in the global shift toward renewable power sources, the need for efficient energy storage solutions becomes increasingly critical. Solar energy is intermittent, meaning it is only available when the sun is shining. To ensure a reliable and constant energy supply, it is essential to store excess energy generated during the day for use during periods of low or no sunlight. Energy storage solutions enable this balance, ensuring that solar energy can be utilized effectively at all times. This section will explore the role of energy storage in solar power systems, including advancements in battery storage technologies, thermal storage in concentrated solar power (CSP) plants, and the integration of hydrogen as an energy storage medium [16].

10.3.1 Need for Energy Storage in Solar Power Systems

One of the main challenges of solar energy is its variability—solar power generation is directly tied to weather conditions and daylight hours. During peak sunlight hours, solar panels often generate more energy than is immediately needed. Without a means to store this excess energy, the surplus is wasted. Conversely, when there is no sunlight, energy demand continues, and the energy stored in the system must be available for use. Energy storage systems are crucial for addressing this issue, as they allow for the smooth integration of solar power into the grid and provide backup during periods of low solar generation. By storing energy during the day and releasing it at night or during cloudy periods, energy storage systems can significantly increase the efficiency and reliability of solar power systems. Stored solar energy can reduce reliance on conventional power plants. In case of grid failure or high energy demand, stored solar energy can provide crucial backup. Energy storage helps balance supply and demand, stabilizing the grid, particularly as solar power becomes a larger portion of the energy mix. Thus, energy storage is indispensable for making solar energy a viable alternative to traditional energy sources on a large scale.

10.3.2 Battery Storage Advancements: Lithium-Ion, Sodium-Ion, and Flow Batteries

The most common form of energy storage for solar power systems is batteries. Over the years, significant advancements have been made in battery technologies, improving energy density, charging speed, cost, and lifespan.

10.3.2.1 Lithium-Ion Batteries

Lithium-ion (Li-ion) batteries are currently the most widely used energy storage solution for residential, commercial, and utility-scale solar power systems. They are favored for their high energy density, long cycle life, and relatively quick charge and discharge rates. Li-ion batteries store and release energy efficiently, making them ideal for balancing the intermittent nature of solar power. Tesla's Powerwall, a widely known home energy storage solution, uses lithium-ion technology. It enables homeowners to store solar energy during the day and use it at night, reducing reliance on the grid [17]. Li-ion batteries can store a large amount of energy in a compact space. These batteries can last for more than ten years, providing a long-term energy storage solution. Lithium-ion batteries are easily scalable, making them suitable for various sizes of installations, from residential to large commercial systems. Li-ion batteries remain relatively expensive, though prices have been steadily decreasing. Lithium and cobalt, key materials for Li-ion batteries, can face supply chain issues and environmental concerns.

10.3.2.2 Sodium-Ion Batteries

Sodium-ion batteries are emerging as a promising alternative to lithium-ion batteries. Sodium is more abundant and less expensive than lithium, which makes sodium-ion batteries a potentially more cost-effective solution for large-scale energy storage. Companies like Natron Energy are developing sodium-ion batteries specifically designed for stationary energy storage applications, including solar energy storage [18]. Sodium is cheaper than lithium, making sodium-ion batteries less expensive to produce. Sodium is widely available, reducing concerns about resource scarcity and geopolitical risks. Sodium-ion batteries are less prone to thermal runaway, making them safer than lithium-ion batteries. Sodium-ion batteries typically have lower energy density compared to lithium-ion batteries, meaning they are bulkier for the same amount of storage. Sodium-ion batteries have slightly lower round-trip efficiency, meaning some energy is lost during the charge-discharge cycle.

10.3.2.3 Flow Batteries

Flow batteries are another innovative energy storage technology that is gaining traction in solar power systems. In a flow battery, energy is stored in liquid electrolyte solutions that flow through the battery during charge and discharge cycles. This design allows for greater scalability, making flow batteries particularly suitable for large-scale solar energy storage. Vanadium redox flow batteries (VRFBs) are commonly used in utility-scale solar applications, such as in the United States and Australia, to store large amounts of solar energy for grid use [19]. Flow batteries can store large amounts of energy by simply increasing the volume of electrolyte solution, making them ideal for utility-scale projects. Flow batteries have a longer operational life than lithium-ion batteries, potentially lasting 20 years or more with minimal degradation. Flow batteries are less likely to overheat and are more stable than conventional batteries. Flow batteries are generally less energy-dense than other types of batteries, requiring more space to store the same amount of energy. The cost of building flow battery systems can be high, though this is expected to decrease as technology advances.

10.3.3 Thermal Storage in CSP Plants

Concentrated solar power (CSP) plants use mirrors or lenses to focus sunlight onto a small area, typically a receiver, to generate heat. This heat is then used to produce electricity, often through steam turbines. One of the significant advantages of CSP systems is their ability to incorporate thermal energy storage, which allows them to produce electricity even when the sun is not shining. Thermal storage in CSP plants is typically achieved using molten salts or other materials that can store heat for extended periods. The heat is stored in these materials during the day and can be used to generate electricity at night or during cloudy periods. The Noor Ouarzazate Solar Complex in Morocco, one of the largest CSP plants in the world, uses molten salt thermal storage to provide up to 6 hours of energy production after the sun has set [20]. Thermal storage can store energy for several hours or even days, providing reliable power even in the absence of sunlight. While CSP systems are capital-intensive, their ability to store energy at scale makes them a viable option for utility-scale energy generation. CSP plants require significant upfront capital for construction and thermal storage systems. CSP systems are most efficient in regions with high direct sunlight, such as deserts, which limits their applicability in cloudy or low-sunlight areas.

10.3.4 Integration of Hydrogen as an Energy Storage Medium

Hydrogen is emerging as a promising energy storage solution for solar power systems. Solar energy can be used to produce hydrogen through the process of electrolysis, where electricity splits water into hydrogen and oxygen. The hydrogen can then be stored and used as a fuel to generate electricity through fuel cells or in other applications. The "Power-to-Gas" project in Germany is an example of using solar energy to produce hydrogen, which is then injected into the natural gas grid, serving as both an energy storage and a method of reducing carbon emissions [21].

Hydrogen can be stored for long periods and transported over large distances, making it ideal for long-term energy storage. Hydrogen has a high energy density by weight, making it suitable for large-scale energy storage. Hydrogen can be used in various sectors, including transportation, industry, and electricity generation, making it a flexible energy carrier. The conversion of electricity to hydrogen and back to electricity (through fuel cells) is less efficient than battery storage. Hydrogen requires specialized storage infrastructure due to its low density and flammability, making it more challenging to manage and transport than other forms of energy storage.

Energy storage is a critical component of solar power systems, enabling the reliable and efficient use of solar energy. With advancements in battery storage technologies, including lithium-ion, sodium-ion, and flow batteries, solar energy storage is becoming more practical and scalable. Thermal storage in CSP plants offers another promising avenue for large-scale energy storage, while hydrogen integration provides a flexible and long-term storage solution. As these technologies continue to evolve, the potential for solar energy to become a primary source of power for the world grows exponentially, contributing to the global effort to reduce carbon emissions and combat climate change.

10.4 SMART SOLAR PLANT MANAGEMENT

The rapid growth of solar power adoption has brought the need for smarter management systems that optimize operations, improve efficiency, and ensure long-term reliability. Smart solar plant management leverages cutting-edge technologies like artificial intelligence (AI), machine learning (ML), real-time monitoring, predictive maintenance, smart grid integration, and blockchain to create a streamlined and effective operational ecosystem. These innovations are transforming traditional solar power plants into intelligent, adaptable systems that maximize energy production while minimizing costs and environmental impacts [22].

10.4.1 Role of Artificial Intelligence (AI) and Machine Learning (ML) in Optimizing Operations

AI and ML are revolutionizing the way solar plants are managed by enabling systems to learn from data, identify patterns, and make informed decisions. These technologies analyze vast amounts of data from sensors, weather forecasts, and performance metrics to optimize energy generation and maintenance schedules. Google's DeepMind uses AI to predict the power output of its solar farms in the United States, improving energy forecasts and grid integration. AI algorithms adjust the angle and orientation of solar panels to track the sun more effectively, maximizing energy capture throughout the day. ML models predict solar power output based on weather data and historical trends, allowing grid operators to balance supply and demand efficiently. AI systems monitor performance data to detect anomalies that may indicate equipment issues, ensuring timely intervention (Table 10.8).

10.4.2 Real-Time Monitoring and Predictive Maintenance

Real-time monitoring systems collect data from various components of a solar plant, such as inverters, panels, and transformers, to provide operators with actionable insights. These systems identify inefficiencies, detect faults, and ensure that every part of the plant operates at its optimal capacity. Traditional maintenance schedules can be inefficient and costly. Predictive maintenance, powered by AI and real-time data analysis, identifies potential equipment failures before they occur. A solar farm in Spain implemented predictive maintenance using IoT sensors and AI analytics, reducing maintenance costs by 25% and improving overall energy production [23].

10.4.3 Integration with Smart Grids

Smart grids are intelligent energy networks that use advanced communication and control technologies to efficiently manage energy supply and demand. Solar power plants integrated with smart grids contribute to a more resilient and adaptable energy system. The Smart Grid Solar Power Project in California integrates multiple solar plants with a regional grid, improving energy reliability and reducing transmission losses. Solar plants

Table 10.8 Benefits of AI Applications

AI Use Case	*Benefit*
Solar Tracking	Increases energy output by up to 20%
Predictive Maintenance	Reduces downtime by 30%
Energy Forecasting	Enhances grid reliability

can dynamically adjust energy output based on real-time demand and supply conditions. Smart grids optimize the use of energy storage systems, ensuring stored solar energy is used during peak demand periods. Solar power plants become part of a decentralized network, allowing energy to be generated, stored, and consumed locally.

10.4.4 Blockchain for Transparent Energy Trading

Blockchain technology is emerging as a game-changer in energy trading by providing a decentralized and transparent platform for transactions. It allows solar plant operators and consumers to trade energy directly without intermediaries, reducing costs and increasing transparency. The Brooklyn Microgrid project in New York uses blockchain to facilitate P2P energy trading among residents with solar panels, creating a localized energy marketplace [24]. Blockchain enables households and businesses with solar panels to sell excess energy directly to others within a network. Solar plants can issue blockchain-based renewable energy certificates (RECs) to prove the origin of their energy, ensuring authenticity and compliance with regulations. Smart contracts automate payments for energy transactions, ensuring timely and accurate settlements.

Smart solar plant management is transforming the renewable energy landscape by combining advanced technologies with innovative solutions. AI and ML optimize operations and maintenance, while real-time monitoring ensures system reliability. Integration with smart grids enhances energy efficiency, and blockchain creates transparent and decentralized energy trading systems. As these innovations continue to evolve, they will play a pivotal role in accelerating the adoption of solar power and creating a sustainable energy future.

10.5 ENVIRONMENTAL AND ECONOMIC CONSIDERATIONS

As the world grapples with climate change and rising energy demands, solar energy emerges as a promising solution that addresses both environmental and economic challenges. However, to maximize its benefits, next-generation solar technologies must tackle concerns related to carbon emissions, ecological impact, panel recycling, affordability, and supportive policies. This chapter explores the multifaceted considerations shaping the adoption and expansion of solar energy systems. Solar energy is inherently clean and renewable, making it a cornerstone in efforts to reduce carbon emissions. Unlike fossil fuels, solar power generation emits no greenhouse gases during operation, contributing significantly to climate change mitigation. A 10 MW solar farm can offset approximately 15,000 metric tons of CO_2 annually, equivalent to removing 3,000 cars from the road [25]. By replacing coal or natural gas power plants, solar systems decrease harmful pollutants such as sulfur dioxide and nitrogen oxides. Solar power uses minimal water compared to conventional thermal power plants, which require substantial water resources for cooling. Large-scale solar farms may lead to habitat disruption if not properly planned. Solutions like floating solar panels and dual-use agrivoltaics address this issue. Though much lower than fossil fuels, manufacturing solar panels generates emissions. Innovations in cleaner production processes are helping to minimize this footprint.

10.5.1 Recycling and Disposal of Solar Panels

With the widespread adoption of solar energy, managing end-of-life panels is becoming increasingly important. Solar panels typically have a lifespan of 25–30 years, after which efficient recycling processes are crucial to minimize environmental impact. The European Union mandates a recycling rate of 85% for end-of-life PV modules under the Waste Electrical and Electronic Equipment (WEEE) Directive. Companies like First Solar have developed closed-loop recycling systems capable of recovering 90% of materials [26]. High-purity silicon in PV cells can be reused in new panels or electronics. About 75% of a solar panel's weight is glass, which can be recycled for various applications. Valuable metals like silver, aluminum, and copper can be extracted and repurposed (Table 10.9).

10.5.2 Cost Trends and Improving Affordability of Solar Technologies

The cost of solar power has plummeted over the past two decades, making it one of the most affordable energy sources. This affordability stems from advancements in technology, economies of scale, and competitive manufacturing. In India, solar tariffs have reached record lows of ₹1.99/kWh, undercutting coal-based power prices [27]. The cost of solar PV modules has dropped by over 80% since 2010 due to improved efficiency and streamlined

Table 10.9 The European Union Mandate Recycling Rate

Material	*Recycling Rate %*	*Application*
Glass	95	New panels or construction
Silicon	85	New solar cells
Metals	90	Electronics and wiring

Table 10.10 Year-Wise Solar Module Price

Year	*Solar Module Price ($/W)*	*Global Installed Capacity (GW)*
2010	2.00	40
2020	0.20	760
2030	0.10	1,500

Table 10.11 Policy and Its Impact

Policy	*Region*	*Impact*
Investment Tax Credit (ITC)	United States	Boosted residential solar installations
Feed-in Tariffs (FiTs)	Germany	Encouraged rapid solar capacity growth
Net Metering	India, Australia	Promoted rooftop solar adoption

production. Innovations in inverters, mounting systems, and energy storage have also lowered installation costs. Solar PV has achieved one of the lowest LCOEs among all energy sources, with utility-scale solar as low as $0.02 per kWh in regions with abundant sunlight (Table 10.10).

10.5.3 Incentives and Policies Driving Next-Generation Solar Adoption

Governments worldwide play a pivotal role in accelerating solar adoption through incentives, subsidies, and supportive regulations. Germany's Renewable Energy Sources Act (EEG) transformed the country into a solar energy leader, achieving over 50 GW of installed capacity by 2023 [28]. FiTs ensure fixed payments for solar-generated electricity fed into the grid, guaranteeing returns for investors. Programs like the U.S. Investment Tax Credit (ITC) offer significant deductions for solar system installations, reducing upfront costs. Net Metering allows consumers to sell excess electricity back to the grid, reducing energy bills. Renewable Portfolio Standards (RPS) mandate a minimum share of renewable energy in the electricity mix, driving utility-scale solar projects (Table 10.11).

Environmental and economic considerations are pivotal to the global transition to solar energy. From minimizing the carbon footprint and recycling solar panels to improving affordability and leveraging supportive policies, the solar energy sector is evolving to meet the challenges of sustainability and cost-effectiveness. By addressing these aspects, next-generation solar technologies will continue to play a transformative role in achieving a cleaner and more equitable energy future.

Table 10.12 Comparison of Ground-Mounted and Floating Solar Plants

Parameter	*Ground-Mounted Solar*	*Floating Solar*
Land Use	High	Minimal
Efficiency	Standard	Higher
Water Conservation	Not Applicable	Significant
Maintenance Costs	Moderate	Higher

10.6 INNOVATIVE PLANT DESIGNS AND APPLICATIONS

As solar technology evolves, innovative designs and applications are emerging to optimize land use, enhance energy generation, and expand solar deployment in diverse environments. These advancements include floating solar farms, agrivoltaics, building-integrated photovoltaics (BIPVs), and solar power plants designed for extreme climates.

10.6.1 Floating Solar Farms: Benefits and Challenges

Floating solar farms, also known as floating photovoltaic systems (FPVs), involve installing solar panels on water bodies like reservoirs, lakes, and dams. This approach minimizes land use and leverages the cooling effects of water to improve efficiency. The Bhadla Floating Solar Plant in India, with a capacity of 600 MW, combines hydropower with FPVs, reducing carbon emissions by 1.2 million tons annually [29]. FPVs are ideal for regions with limited land availability; they free up land for other purposes. Water cooling reduces panel temperatures, boosting energy output by 5–15%. Panels provide shade, lowering evaporation rates from reservoirs and conserving water resources. Floating solar farms can be integrated with hydropower plants to enhance grid stability. FPV installation and anchoring systems in water bodies are costlier than ground-mounted systems. They have potential impacts on aquatic ecosystems, such as reduced sunlight penetration. Corrosion and biofouling can increase operational challenges [30].

10.6.2 Agrivoltaics: Combining Agriculture with Solar Energy

Agrivoltaics integrates solar panels with agricultural land, allowing crops and solar energy production to coexist. This synergy benefits farmers and energy producers alike. It maximizes the productivity of agricultural lands, especially in areas with limited resources. Panels provide partial shade, reducing heat stress on crops and conserving soil moisture. Reflected light from crops can increase the efficiency of bifacial solar panels. Ensuring adequate sunlight for crops and optimal panel orientation requires meticulous planning. A German agrivoltaics project successfully grew tomatoes and lettuce under solar panels, reducing irrigation needs by 30%. Initial setup costs for agrivoltaics systems are higher due to specialized structures. India's KUSUM scheme promotes solar pumps for irrigation, enabling farmers to sell excess electricity to the grid.

10.6.3 Building-Integrated Photovoltaics (BIPVs)

BIPVs incorporate solar panels directly into building materials, such as rooftops, facades, and windows, transforming structures into energy-generating units. BIPV panels blend seamlessly with architectural designs, enhancing the visual appeal of buildings. They eliminate the need for separate solar installations, making them ideal for urban areas. BIPVs reduce electricity bills and contribute to building energy efficiency. Panels can act as an additional layer, reducing heat transfer and cooling costs.

Customization and installation costs of BIPVs are higher than traditional solar panels. The Edge in Amsterdam, the world's greenest building, uses BIPVs on its facade to generate a significant portion of its energy needs. Aesthetic considerations may compromise optimal panel orientation and efficiency. Tesla Solar Roof offers homeowners an innovative way to combine durability, aesthetics, and solar energy generation.

10.6.4 Solar Power Plants in Extreme Climates

Extreme climates, such as deserts, polar regions, and high-altitude areas, pose unique challenges and opportunities for solar power plants. Deserts are ideal locations for large-scale solar power plants due to their high solar irradiance, which ensures maximum energy capture. However, challenges such as dust accumulation can reduce panel efficiency, and extreme temperatures can strain system components. An example is the Noor Solar Complex in Morocco, which utilizes concentrated solar power (CSP) technology to generate energy even after sunset through efficient thermal storage systems. Polar regions offer long summer days, providing extended sunlight hours that are perfect for solar energy production. However, winter months require energy storage to ensure a continuous power supply. Bifacial panels are effective in such regions, as they capture reflected light from snow, enhancing energy generation. A pilot solar project in Antarctica powers research stations during the summer months, highlighting the viability of solar energy in extreme climates. High-altitude regions benefit from increased solar intensity and cooler temperatures, both of which enhance the efficiency of solar panels. However, challenges include strong winds and the effects of UV radiation, which can degrade materials over time. The 1 GW solar plant in Tibet, located above 4,000 meters, operates efficiently despite these challenges, demonstrating the potential for solar power at high altitudes [31–33].

Innovative plant designs such as floating solar farms, agrivoltaics, BIPVs, and solutions for extreme climates demonstrate the versatility of solar energy technologies. These designs enhance energy generation while addressing challenges related to land use, environmental compatibility, and integration into existing infrastructures. By continuing to explore and implement these advancements, solar energy can become a more accessible and sustainable solution for diverse global energy needs.

10.7 CASE STUDIES OF NEXT-GENERATION SOLAR PLANTS

Next-generation solar plants exemplify the potential of cutting-edge technology, innovative designs, and advanced management practices in optimizing solar energy production. This section delves into real-world examples of advanced solar installations, their unique features, and the lessons learned to inform future projects.

10.7.1 Noor Ouarzazate Solar Complex, Morocco

The Noor Ouarzazate Solar Complex is one of the world's largest concentrated solar power (CSP) projects, integrating advanced technologies for maximum efficiency. The complex consists of four phases: Noor I, II, III, and Noor IV, combining CSP and photovoltaic (PV) technologies [31].

- **Technology Used:** Parabolic troughs (Noor I & II), solar towers (Noor III), and PV panels (Noor IV).
- **Energy Storage:** Molten salt storage allows energy production for 3–8 hours post-sunset.
- **Capacity:** Over 580 MW, supplying clean energy to 1.1 million people.

By utilizing solar power, large-scale solar plants significantly decrease the need for fossil fuels, resulting in a substantial reduction in carbon emissions. For example, a major solar installation can cut annual CO_2 emissions by approximately 760,000 tons, helping mitigate the impacts of climate change [32]. A hybrid setup combining concentrated solar power (CSP) and photovoltaic (PV) technologies has demonstrated its effectiveness. This approach ensures a more consistent and reliable energy supply, as CSP can store energy for night-time use while PV systems perform optimally during daylight hours, enhancing the overall energy generation capacity.

10.7.2 Tengger Desert Solar Park, China

Known as the "Great Wall of Solar," the Tengger Desert Solar Park is one of the largest PV installations globally, covering 43 km^2 in the Tengger Desert.

- **Capacity:** 1,547 MW of installed capacity.
- **Design:** The park uses fixed-mount PV panels to reduce maintenance costs in a desert environment.

China has utilized its expansive desert regions for large-scale solar power installations, minimizing environmental impact. These areas, with their high solar irradiance, are ideal for generating vast amounts of energy without disturbing populated ecosystems or agricultural land. China's investments in solar energy have been pivotal in meeting the country's ambitious renewable energy goals. Solar power plants in these desert regions significantly contribute to China's overall renewable energy capacity, advancing the country's transition to cleaner energy sources (Table 10.13).

10.7.3 Bhadla Solar Park, India

Bhadla Solar Park in Rajasthan is the largest operational solar park in the world, located in one of India's sunniest regions.

- **Capacity:** 2,245 MW.
- **Technology:** Combination of monocrystalline and polycrystalline PV panels.
- **Innovative Features:** Smart monitoring systems optimize panel orientation and output.

Solar power installations, particularly large-scale projects, provide affordable and reliable electricity to millions of households, especially in regions that previously lacked access to consistent power. This has helped improve

Table 10.13 Features of Tengger Solar Park

Features	*Tengger Solar Park*
Land Area	43 km^2
Capacity	1,547 MW
Technology	Fixed-mount PV panels
Annual CO_2 savings	~2.6 million tons

the quality of life and promote economic development in underserved areas. The growth of renewable energy, especially solar, highlights the effectiveness of public-private partnerships. By combining government incentives with private sector investment and expertise, these collaborations have accelerated the development and deployment of solar power systems, making renewable energy more accessible and scalable.

10.7.4 Agua Caliente Solar Project, United States

The Agua Caliente Solar Project in Arizona leverages thin-film PV technology for efficient energy generation under high-temperature conditions.

- **Capacity:** 290 MW.
- **Technology:** Thin-film cadmium telluride (CdTe) PV modules.
- **Environmental Adaptation:** Panels are designed to withstand the harsh desert climate.

By strategically placing solar panels in areas with minimal environmental disruption, such as rooftops, barren land, or unused industrial zones, solar installations can reduce the impact on natural ecosystems. This optimized placement ensures energy production while preserving surrounding habitats and reducing land usage. Large-scale solar projects are designed to meet substantial energy needs. For example, a well-optimized solar power plant can generate enough electricity to power over 100,000 homes, demonstrating the potential of solar energy to support entire communities sustainably [33].

10.8 CHALLENGES AND FUTURE DIRECTIONS

As solar energy becomes an increasingly critical component of global energy strategies, several challenges must be addressed to fully realize its potential. This section examines the major hurdles in solar technology development and adoption, as well as promising future directions that could shape the next decade [34–37].

The efficiency of solar panels and other solar technologies remains a key challenge. While commercial silicon-based photovoltaic (PV) panels typically achieve efficiencies between 15% and 22%, they lose significant energy due to heat, shading, and conversion inefficiencies (Table 10.14).

Investing in research and development (R&D) to improve manufacturing processes and material usage efficiency will help bridge these gaps. For instance, recyclable materials and sustainable mining practices could ensure long-term scalability. Solar adoption varies significantly across regions due to disparities in economic capacity, technological access, and geographic

Table 10.14 Efficiency of Solar Panels

Technology	*Efficiency Range*	*Key Challenge*
Monocrystalline PV	18%–22%	High cost of production
Perovskite Solar Cells	25%–30% (lab scale)	Stability and scalability
Tandem Solar Cells	30%–35% (potential)	Complexity of manufacturing

conditions [38–41]. Developed nations like Germany and the United States have made substantial progress in solar adoption, supported by policies and incentives. Developing nations in Africa and South Asia lag behind due to financial and technical constraints. Targeted funding and international collaboration can help bridge global adoption disparities. Initiatives like the International Solar Alliance (ISA) aim to support solar adoption in underdeveloped regions [42–44].

Future research will focus on developing materials that enhance performance, durability, and cost-efficiency. Solid-state batteries are emerging as a promising technology with significantly higher energy density compared to traditional lithium-ion batteries. Their compact design and safety benefits make them an ideal choice for large-scale energy storage, offering the potential for longer-lasting, more efficient solar power systems [45–47]. Hydrogen holds great potential as both energy storage medium and fuel. It can store excess energy generated by solar power and be converted into electricity when needed, or used directly as fuel in various sectors, contributing to a more sustainable and versatile energy infrastructure. Artificial intelligence (AI) and machine learning (ML) can optimize solar plant performance by analyzing real-time data. Predictive maintenance powered by AI reduces downtime and operational costs [48–50].

Addressing the challenges of efficiency, scalability, and adoption disparities is crucial to the success of next-generation solar energy. By investing in R&D, fostering international cooperation, and aligning policies with global sustainability goals, solar technology can become the backbone of the global energy transition. Future innovations in materials, energy storage, and AI-driven management systems hold the promise of unlocking solar energy's full potential, paving the way for a sustainable and equitable energy future [51].

REFERENCES

[1] Green, M. A.; Bremner, S. P.; Wenham, S. R. "Photovoltaic Technology: A Review of Recent Advances," *Progress in Photovoltaics: Research and Applications*, 2017, 25 (12), 1102–1119.

[2] REN21. *Renewable Energy Policy Network for the 21st Century: Renewables 2020 Global Status Report*. REN21 Secretariat, 2020.

[3] Green, M. A.; Emery, K.; Hishikawa, Y.; Warta, W.; Dunlop, E. D. "Solar Cell Efficiency Tables (Version 56)," *Progress in Photovoltaics: Research and Applications*, 2020, 28 (1), 3–15.

[4] Green, M. A.; Emery, K.; Hishikawa, Y.; Warta, W.; Dunlop, E. D. "Solar Cell Efficiency Tables (Version 56)," *Progress in Photovoltaics: Research and Applications*, 2020, 28 (1), 3–15.

[5] Miyatake, K.; Nakanishi, K.; Matsui, T.; Takahashi, M.; Fukui, T. "Recent Advances in Thin-Film Photovoltaic Solar Cells," *Journal of Photovoltaics*, 2017, 7 (4), 1–8.

[6] Ball, J. M.; Lee, M. M.; Hey, A.; Snaith, H. J. "Solvent-Enhanced Crystallization of Perovskite for High Efficiency Solar Cells," *Nature Photonics*, 2013, 8 (3), 151–156.

[7] Kane, A. F.; Hoke, E. T.; Nguyen, W. H.; Beard, M. C.; Snaith, H. J.; McGehee, M. D. "High-Efficiency Tandem Solar Cells," *Nature Materials*, 2022, 21 (8), 1071–1078.

[8] Dersch, J.; Zeng, F.; Sattler, C.; Kreutz, A.; Sopha, J. "Advancements in Concentrated Solar Power (CSP) Technologies: Efficiency, Scalability, and Storage," *Renewable and Sustainable Energy Reviews*, 2021, 141, 110848.

[9] Romero, M.; Valle, L. "Parabolic Trough Solar Power: Advancements and Applications," *Energy Procedia*, 2013, 49, 273–282.

[10] Behar, O.; Garcia, C.; Faiman, D. "Solar Tower Systems: Efficiency, Challenges, and Future Prospects," *Renewable and Sustainable Energy Reviews*, 2016, 58, 145–156.

[11] O'Neill, S.; Foster, E.; McLean, J. "Fresnel Reflector Systems in Concentrated Solar Power: A Review of Performance and Applications," *Energy Procedia*, 2015, 69, 1420–1428.

[12] El-Amin, M.; Morsi, M.; Sayed, H. "Hybrid CSP-PV Systems for Enhanced Solar Power Generation: Case Study of the Noor Midelt Project," *Renewable Energy*, 2020, 145, 1989–1998.

[13] Law, M.; Moore, L. E. C.; Alivisatos, P. M. "Quantum Dot Solar Cells," *Nano Letters*, 2005, 5 (2), 160–164.

[14] Tang, C. W.; Van Slyke, S. A.; Chen, C. H. "Organic Electroluminescent Diodes," *Applied Physics Letters*, 1987, 51 (12), 913–915.

[15] Xu, H.; Xu, X.; Zhang, L.; et al. "Transparent Solar Cells for Building-Integrated Photovoltaics," *Nature Energy*, 2020, 5 (10), 903–910.

[16] Liu, Y.; Wang, C.; Zhang, L.; et al. "Advancements in Energy Storage for Solar Power Systems: Battery Storage, Thermal Storage, and Hydrogen Integration," *Renewable and Sustainable Energy Reviews*, 2021, 138, 110563.

[17] Nagaura, T.; Tozawa, K.; Watanabe, M.; et al. "Lithium-Ion Batteries: Fundamentals and Applications in Solar Energy Storage," *Journal of Power Sources*, 2020, 450, 227674.

[18] Tarascon, J. M.; Armand, M. "Sodium-Ion Batteries: A New Era of Energy Storage," *Nature Materials*, 2021, 20, 1155–1163.

[19] Skyllas-Kazacos, M.; Liu, H.; Chakrabarti, M.; Ker, P. J. "Vanadium Redox Flow Batteries: A Review of Developments and Applications," *Energy Storage Materials*, 2021, 34, 9–29.

[20] Benali, M.; Kousksou, T.; Jamil, A.; Laqbaqbi, M.; El Rhafiki, T. "Concentrated Solar Power (CSP) and Thermal Storage: A Review of Technologies and Applications," *Renewable and Sustainable Energy Reviews*, 2019, 104, 257–273.
[21] Zeng, K.; Zhang, L.; Xu, B.; Li, S. "Hydrogen Production Via Water Electrolysis: A Review of Recent Advances," *Renewable and Sustainable Energy Reviews*, 2015, 42, 67–85.
[22] Zeng, K.; Zhang, L.; Xu, B.; Li, S. "Hydrogen Production Via Water Electrolysis: A Review of Recent Advances," *Renewable and Sustainable Energy Reviews*, 2015, 42, 67–85.
[23] By Addressing Issues Proactively, Solar Plants Can: Zeng, K.; Zhang, L.; Xu, B.; Li, S. "Hydrogen Production Via Water Electrolysis: A Review of Recent Advances," *Renewable and Sustainable Energy Reviews*, 2015, 42, 67–85.
[24] Zeng, K.; Zhang, L.; Xu, B.; Li, S. "Hydrogen Production Via Water Electrolysis: A Review of Recent Advances," *Renewable and Sustainable Energy Reviews*, 2015, 42, 67–85.
[25] International Energy Agency (IEA). *World Energy Outlook 2020*. Paris: IEA, 2020.
[26] European Commission. *Directive 2012/19/EU on Waste Electrical and Electronic Equipment (WEEE)*. Official Journal of the European Union, 2012.
[27] International Renewable Energy Agency (IRENA). *The Changing Role of Solar Power in Energy Transition*. IRENA, 2023.
[28] Federal Ministry for Economic Affairs and Energy (BMWi). *Germany's Renewable Energy Sources Act (EEG): Achievements and Future Plans*. Federal Ministry for Economic Affairs and Energy, 2023.
[29] Bhadla Floating Solar Plant. *Bhadla Floating Solar Plant: An Innovative Approach to Clean Energy*. Bhadla Solar, 2023.
[30] Tibet 1 GW Solar Plant. *Harnessing Solar Power at High Altitudes: The 1 GW Solar Plant in Tibet*. Tibet Solar Energy, 2023.
[31] Noor Ouarzazate Solar Complex. *Noor Ouarzazate Solar Complex: A Landmark CSP and PV Integration*. Noor Ouarzazate Solar Project, 2023.
[32] Turney, D.; Fthenakis, V. Environmental Impacts from the Installation and Operation of Large-Scale Solar Power Plants," *Renewable and Sustainable Energy Reviews*, 2011, 15(6), 3261–3270.
[33] Mendelsohn, M.; Lowder, T.; Canavan, B. *Utility-Scale Concentrating Solar Power and Photovoltaic Projects: A Technology and Market Overview* (No. NREL/TP-6A20-51137). National Renewable Energy Lab (NREL), Golden, CO (United States), 2012.
[34] Akimov, L.; Lvov, V.; de Martino di Montegiordano, D.; De Mei, K.; Osipov, N.; Ostrovaia, A.; . . . Terleev, V. "Shading System Design and Solar Gains Control for Buildings Passive Energy-Efficiency Improvement," In *Technological Advancements in Construction: Selected Papers* (pp. 13–24). Springer International Publishing, 2021.
[35] Kundu, A.; Kumar, A.; Awasthi, M. K.; Dwivedi, G.; Shukla, A. K. "An Explorative Study of Earth-Air Heat Exchanger for Sustainable Building Applications," *International Journal of Energy for a Clean Environment*, 2025, 26 (3), 1–22. https://doi.org/10.1615/InterJEnerCleanEnv.2024052513.

[36] Razaviyn, Z.; Heidari, M.; Thangavel, S.; Verma, V.; Kumar, A.; Yadav, A. K. "Marine Energy Harvesting From Fluid Flow Via Vortex Induced Vibrations," *International Journal of Thermofluids*, 2025, 25, 101015. https://doi.org/10.1016/j.ijft.2024.101015.

[37] Awasthi, M. K.; Shukla, A. K.; Kumar, A.; Dutt, N.; Yadav, D. "Instability of Casson Fluid–Viscous Fluid Interfaces," In *Advances in Computational Methods and Modeling for Science and Engineering* (pp. 113–121). Morgan Kaufmann, 2025. https://doi.org/10.1016/B978-0-44-330012-7.00017-5.

[38] Singh, B.; Kaur, P.; Yadav, A. K.; Awasthi, M. K.; Kumar, A. "Chapter 11: Design and Analysis of Solar Tracking System for PV Thermal Performance Enhancement," In *Heat Transfer Enhancement Techniques: Thermal Performance, Optimization and Applications*, A. Kumar, N. Dutt, M. K. Awasthi, Editors (pp. 251–267). John Wiley & Sons, Inc. USA, 2024. https://doi.org/10.1002/9781394270996.ch11.

[39] Kesharvani, S.; Katre, S.; Banait, S. K.; Dwivedi, G.; Kumar, A. "Chapter 1: Fueling the Future: Sustainable Biodiesel Production Strategies for a Cleaner Tomorrow," In *Clean Energy: Technology, Advances, and Applications*, T. N. Verma, T. S. Sing, U. Rajak, P. Nashine, G. Dwivedi, A. Kumar, Editors (pp. 1–26). Boca Raton, FL: CRC Press, 2024. https://doi.org/10.1201/9781003521341–1.

[40] Kumar, A. "Chapter 11: Evolution of Carbon Capture, Utilization and Storage Technology: Past, Present and Future Developments," In *Clean Energy: Technology, Advances, and Applications*, T. N. Verma, T. S. Sing, U. Rajak, P. Nashine, G. Dwivedi, A. Kumar, Editors (pp. 154–171). Boca Raton, FL: CRC Press, 2024. https://doi.org/10.1201/9781003521341–11.

[41] Kumar, A.; Gangwar, A. K. S.; Patra, B. B.; Dwivedi, G. "Chapter 12: Optimizing Solar Air Heater Performance: A Comparative Analysis of Artificial Roughness Geometries," In *Clean Energy: Technology, Advances, and Applications*, T. N. Verma, T. S. Sing, U. Rajak, P. Nashine, G. Dwivedi, A. Kumar, Editors (pp. 172–186). Boca Raton, FL: CRC Press, 2024. https://doi.org/10.1201/9781003521341–12.

[42] Kumar, A. "Chapter 15: Solar-Integrated Hydrogen Production and Utilization in India," In *Clean Energy: Technology, Advances, and Applications*, T. N. Verma, T. S. Sing, U. Rajak, P. Nashine, G. Dwivedi, A. Kumar, Editors (pp. 243–258). Boca Raton, FL: CRC Press, 2024. https://doi.org/10.1201/9781003521341–15.

[43] Prakash, V.; Saraswat, V.; Dewangan, A.; Yadav, A. K.; Kumar, A. "Chapter 16: Production and Utilization of Hydrogen as a Future Fuel for Electric Vehicles: A Review," In *Clean Energy: Technology, Advances, and Applications*, T. N. Verma, T. S. Sing, U. Rajak, P. Nashine, G. Dwivedi, A. Kumar, Editors (pp. 259–281). Boca Raton, FL: CRC Press, 2024. https://doi.org/10.1201/9781003521341–16.

[44] Kumar, A. " Chapter 19: Building Resilience to Airbone Pathogens: An Indoor Air Quality Model," In *Clean Energy: Technology, Advances, and Applications*, T. N. Verma, T. S. Sing, U. Rajak, P. Nashine, G. Dwivedi, A. Kumar, Editors (pp. 318–333). Boca Raton, FL: CRC Press, 2024. https://doi.org/10.1201/9781003521341–19.

[45] Awasthi, K. M.; Dutt, N.; Kumar, A.; Yadav, D. "Chapter 1: Nanofluid Dynamics Properties and Applications," In *Nanofluid Dynamics and Transport*

Phenomenon, M. K. Awasthi, Editor (pp. 1–20). Boca Raton, FL: CRC Press, 2024. https://doi.org/10.1201/9781003527251-1.

[46] Kumar, A.; Awasthi, K. M.; Dutt, N. "Chapter 13: Nanofluidic Applications in Sustainable Food Technology and Agriculture," In *Nanofluid Dynamics and Transport Phenomenon*, M. K. Awasthi, Editor (pp. 224–242). Boca Raton, FL: CRC Press. https://doi.org/10.1201/9781003527251-13.

[47] Pimoli, D. S.; Tomar, P.; Kumar, A.; Dwivedi, G.; Shukla, A. "Experimental Investigation of Friction Factor and Heat Transfer Enhancement Using Circular Perforated Twisted Tape Inserts in Heat Exchangers," *International Journal of Energy for a Clean Environment*, 26 (3). https://doi.org/10.1615/InterJEnerCleanEnv.2024053616.

[48] Mokhtari, M.; Jamei, S.; Razaviyan, Z.; Heidari, M.; Thangavel, S.; Kumar, A. "Numerical and Hydrodynamic Analysis of a Container Vessel Due to the Change of Collision Wave Angle at Different Speeds in Regular Wave," *Journal of Mechanical Engineering and Sciences*, 2024, 18 (4), 10247–10260. https://doi.org/10.15282/jmes.18.4.2024.3.0809.

[49] Kumar, A.; Agrawal, P. S.; Singla, Y. K. *Sustainable Development With Carbon Capture, Utilization and Storage Technology*. Switzerland: Springer Nature, 2025.

[50] Yadav, D.; Awasthi, M. K.; Kumar, A. *Next Generation Renewable Thermal Energy Harvesting, Conversion & Storage Technologies*. Elsevier, 2025.

[51] Kumar, A.; Dutt, N.; Awasthi, M. K., Editors. *Heat Transfer Enhancement Techniques: Thermal Performance, Optimization and Applications*. John Wiley & Sons, Inc., 2024. https://doi.org/10.1002/9781394270996.

Chapter 11

Novel Hybrid Materials for Next-Generation Sustainable Solar Cells

Aviral Chaurasia, and Saurabh Mishra

11.1 INTRODUCTION

The global energy landscape is undergoing a dramatic transformation, driven by the imperative to meet escalating energy demands while simultaneously mitigating the adverse effects of climate change (Tamiaki & Kichishima, 2025; Hassan *et al.*, 2024; Albert, 2021). This dual challenge necessitates a rapid transition toward renewable energy sources, which offer a sustainable and environmentally friendly alternative to conventional fossil fuels (Singh *et al.*, 2024; Adelekan *et al.*, 2024; Kalair *et al.*, 2021). Among the various renewable energy options, solar energy stands out as a particularly promising candidate due to its abundance, sustainability, and potential to provide clean and inexhaustible power (Pimoli *et al.*, 2025; Panwar *et al.*, 2011; Dey *et al.*, 2022). However, widespread adoption of solar energy technologies hinges on overcoming the limitations of current photovoltaic systems, particularly the high cost and environmental impact associated with traditional silicon-based solar cells (Karafillidis, 2022, Panagoda *et al.*, 2023; Tiwari *et al.*, 2022; Tamrakar *et al.*, 2024). This has spurred extensive research into alternative materials and technologies, including organic photovoltaics, which offer advantages in terms of cost, flexibility, and ease of fabrication. Despite these advantages, OPVs still face significant challenges related to efficiency and stability, necessitating further innovation to unlock their full potential (Fuziki *et al.*, 2023). Natural organic compounds, particularly chlorophyll, have attracted considerable attention for their remarkable light-harvesting capabilities and potential applications in solar energy conversion (Tamiaki & Kichishima, 2025; Saxena *et al.*, 2022; Prakash & Janarthanan, 2023; Duan *et al.*, 2020).

Chlorophyll, the green pigment responsible for photosynthesis in plants, is found in the chloroplasts of green plants. The basic structure of a chlorophyll molecule is a porphyrin ring, coordinated to a central atom. Magnesium (Mg) exhibits strong absorption in the visible region of the solar spectrum, making it an attractive candidate for photovoltaic applications (Karafillidis, 2022; Kumara *et al.*, 2017; Koifman *et al.*, 2021; Roslan *et al.*, 2018).

DOI: 10.1201/9781003634737-11

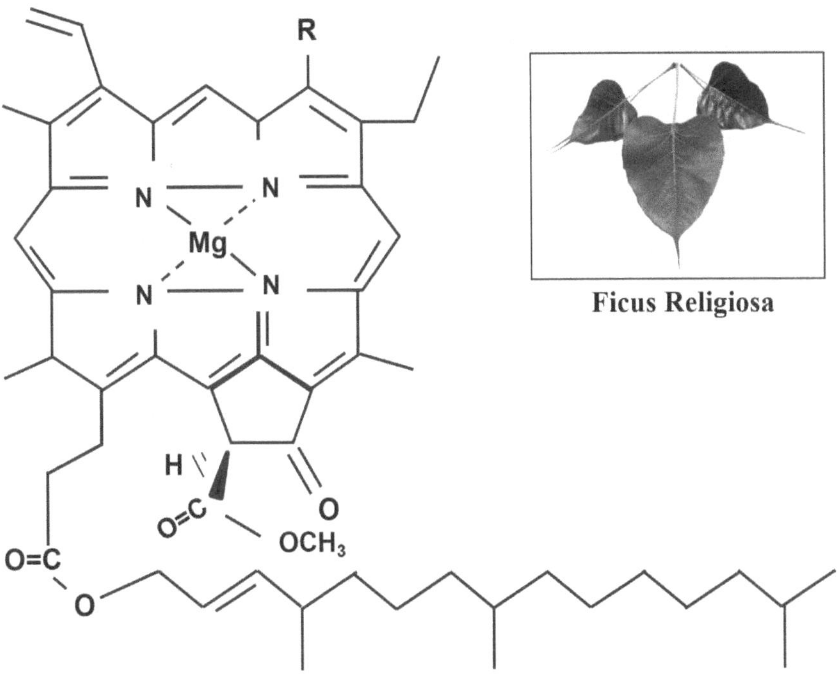

Figure 11.1 Ficus religiosa and molecular structure of chlorophyll.

Figure 11.1 represents the molecular structure of chlorophyll, with the central magnesium ion (Mg) coordinated in the porphyrin ring and the phytol tail extending from the structure. The groups attached to the porphyrin ring, such as the carbonyl group (C = O) and methoxy group (OCH3), further define its structure, making it essential for its light-absorbing properties in photosynthesis.

The molecular structure of chlorophyll facilitates efficient light absorption and energy transfer. However, the practical application of chlorophyll in solar cells is hampered by its inherent instability and low charge carrier mobility (Buscemi *et al.,* 2022; Iman *et al.,* 2024; Musazade *et al.,* 2018; Meng *et al.,* 2022). To overcome these limitations, researchers have explored the integration of nanoparticles with chlorophyll, aiming to enhance its optical and electronic properties and improve its performance in photovoltaic devices (Arof & Ping, 2017; Koifman *et al.,* 2021; Rekioua *et al.,* 2023).

Chlorophyll has been extensively studied for its potential in photosynthetic solar cells, offering a sustainable and eco-friendly alternative to conventional photovoltaic materials (Hoerner, 2013; Mariotti *et al.,* 2020; Singh and Mukherjee, 2022; Dhorkule *et al.,* 2024). Its ability to efficiently absorb light in the visible spectrum and convert it into electrical energy makes it a promising candidate for next-generation solar cells (Koifman *et al.*, 2021). However, the inherent limitations of chlorophyll, such as low charge carrier mobility and photostability, have led researchers to explore hybrid systems that combine chlorophyll with advanced nanomaterials (Chen *et al.*, 2005; Shan *et al.*, 2024; Liang *et al.*, 2024). These hybrid systems aim to leverage the unique properties of both chlorophyll and nanomaterials to create synergistic effects that enhance the overall performance of solar cells (Njeru *et al.*, 2023; Xiong *et al.*, 2023). Previous research has demonstrated that doping chlorophyll with nanoparticles, such as titanium dioxide and copper dioxide, can significantly alter its electronic, optical, photochemical, and photocatalytic properties, opening new possibilities for various applications (Chen *et al.*, 2005). The incorporation of foreign nanomaterials into the atomic structure of chlorophyll can enhance light absorption, facilitate efficient charge separation, and transport, and improve the overall stability of the material (Koifman *et al.*, 2021; Mony *et al.*, 2022). These improvements address key challenges faced by chlorophyll-based solar cells and pave the way for the development of more efficient and sustainable photovoltaic devices. Titanium dioxide nanoparticles have emerged as a promising material for solar energy applications due to their high stability, wide bandgap, and excellent charge transport properties (Fuziki *et al.*, 2023; Chandoliya *et al.*, 2024). They have been widely used in dye-sensitized solar cells (DSSCs), where they serve as the electron transport layer (Omar and Abdullah, 2014; Sugathan *et al.*, 2015). Recent studies have shown that doping TiO_2 nanoparticles into organic matrices can significantly improve the efficiency and stability of photovoltaic devices (O'Regan & Grätzel, 1991). Guo and Ma (2014) focused on nanostructured nitrogen doping of TiO_2 for use in DSSCs, while Amao (2008) had examined photoenergy conversion systems based on photosynthesis dyes conjugated with nanoparticles. Yang *et al.* (2025) had investigated the photocatalytic reduction of Cr using TiO_2 templated and confined with chlorophyll. Moreover, Zhan *et al.* (2019) provided insights into the engineering of highly efficient porous semiconductor photocatalysts derived from metal-organic frameworks, which could be relevant to the development of advanced materials for solar energy conversion. These practices have led to increased interest in exploring the potential of TiO_2-doped chlorophyll as a novel material for solar energy conversion.

In this research study, we investigate the synthesis and characterization of TiO_2-doped chlorophyll and its potential application in solar energy conversion. We employed a simple and cost-effective method to incorporate TiO_2 nanoparticles into the chlorophyll structure, creating a hybrid material with enhanced properties. The optical properties of the TiO_2-doped chlorophyll were investigated using UV-Visible spectroscopy to assess its light absorption characteristics, photon energy levels, and bandgap. The findings of this study provide valuable insights into the potential of TiO_2-doped chlorophyll as a sustainable and efficient material for solar energy applications.

11.2 MATERIALS AND METHODOLOGY

11.2.1 Materials

Fresh leaves of *Ficus religiosa* were harvested for chlorophyll extraction due to their readily available source of chlorophyll. Ethanol (99.9% purity, procured from Byahust Scientico, Jaipur, Rajasthan, India) was selected as the extraction solvent due to its efficacy in disrupting the cell walls of Ficus religiosa leaf tissue and solubilizing chlorophyll. Titanium dioxide (TiO_2) nanoparticles were procured from Sigma-Aldrich Chemicals Private Limited, Bangalore, India.

11.2.2 Methodology

11.2.2.1 Preparation of Chlorophyll Extract

Chlorophyll was extracted from fresh *Ficus religiosa* leaves using ethanol as demonstrated by Putra *et al.* (2017). The leaves were first cleaned with distilled water to remove any surface contaminants. Then, after washing, the leaves were cut into small pieces to increase the surface area for efficient solvent extraction. The leaves were dried in a hot air oven at 60°C for 1 hour to remove moisture and facilitate grinding. The dried leaves were then ground into a fine powder using a mortar and pestle as shown in Figure 11.2 (a).

For chlorophyll extraction, approximately 10 grams of the powdered leaves were mixed with 100 mL of ethanol as the solvent. This grinding process served to disrupt the cell walls of the leaf tissue, releasing the chlorophyll and facilitating its dissolution into the ethanol solvent. The mixture was stirred continuously at room temperature for 1 hour to ensure complete extraction of chlorophyll, as shown in Figure 11.2 (b). The solution was then filtered using Whatman filter paper (No.1) to separate the chlorophyll-ethanol solution (filtrate) from the solid leaf residue. Finally, the filtrate containing the extracted chlorophyll was stored in a clean, airtight glass container to prevent contamination.

Figure 11.2 **(a)** Dried leaves process.

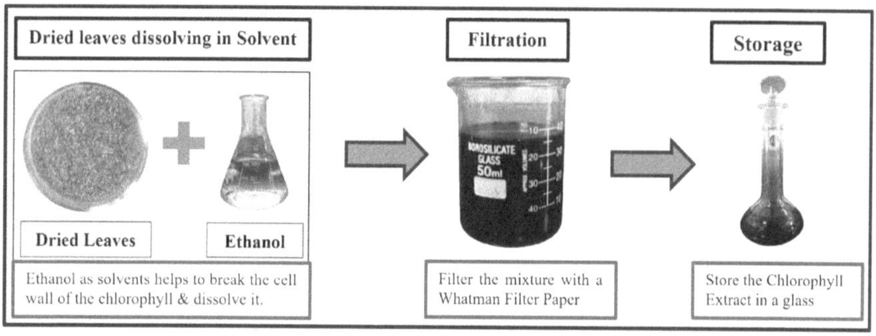

Figure 11.2 **(b)** Chlorophyll extraction process.

11.2.2.2 Demetalation of Chlorophyll and Doping with TiO_2

To facilitate the binding of TiO_2 nanoparticles, the magnesium ion at the center of the chlorophyll molecule, as shown in Figure 11.1, was removed via acid treatment (Harris *et al.*, 2020), converting chlorophyll to pheophytin. This process is consistent with methods described in previous studies (Selan *et al.*, 2021; Arifin *et al.*, 2017). The demetalation was achieved by titrating the chlorophyll solution with 0.1 M hydrochloric acid, prepared by diluting concentrated HCl with distilled water. The titration was performed under continuous stirring, and the demetallization of chlorophyll was indicated by a distinct color change from bright green to olive green or brown. This color change is a well-documented indicator of the removal of the Mg^{2+} ion from the porphyrin ring (Benkerroum, 2019).

Subsequently, the solution was neutralized to a pH of 7–8 using 0.1 M sodium hydroxide. The resulting pheophytin was then separated and collected by centrifugation. The successful removal of the magnesium ion and the formation of pheophytin were confirmed through spectrophotometric analysis, as described in earlier research (Arifin *et al.*, 2017).

A TiO_2 nanoparticle suspension was prepared by dispersing TiO_2 nanoparticles in ethanol at a concentration of 50–100 mg/mL and subjecting the mixture to ultrasonication for 30 minutes to ensure uniform dispersion. This method is consistent with nanoparticle preparation techniques described in previous studies (Harris *et al.*, 2020; Selan *et al.*, 2021). The nanoparticle suspension was then added to the solution containing Mg-free chlorophyll (pheophytin). The mixture was added dropwise and heated to 40–50°C for 2–3 hours to promote interaction and binding between the TiO_2 nanoparticles and the porphyrin ring of the pheophytin molecule. This step is critical for ensuring effective doping, as highlighted in earlier research (Arifin *et al.*, 2017).

Following the heating step, the TiO_2-doped chlorophyll product was collected by centrifugation at 5,000 rpm for 10–15 minutes. The collected

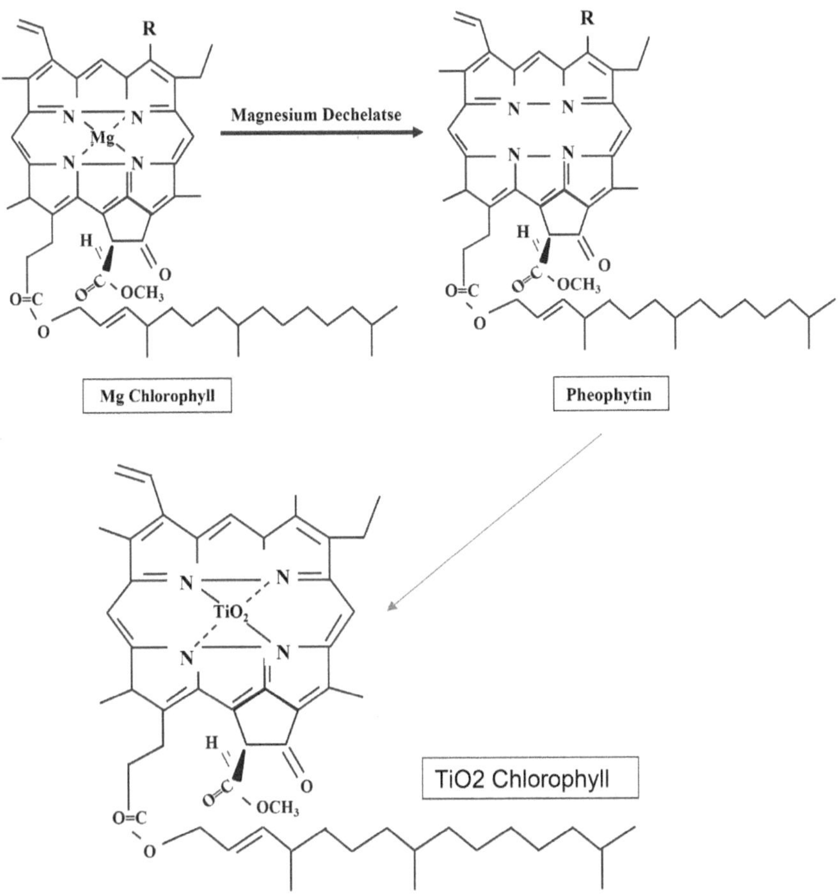

Figure 11.3 Mg^{2+} Substitution with TiO_2 reaction mechanism (Selan *et al.*, 2021; Arifin *et al.*, 2017).

material was washed with ethanol to remove any unbound TiO_2 nanoparticles and residual by-products from the synthesis process. Finally, the TiO_2-doped chlorophyll was dried at a temperature of 40–50°C. The successful binding of TiO_2 nanoparticles to the pheophytin structure was confirmed through spectroscopic analysis, as described in prior work (Benkerroum, 2019).

11.2.2.3 Characterization Techniques

The optical properties, light absorption, bandgap, and photon energy level of the TiO_2-doped chlorophyll, were evaluated with UV-Visible absorption spectroscopy. A Shimadzu UV-2600 spectrophotometer was employed

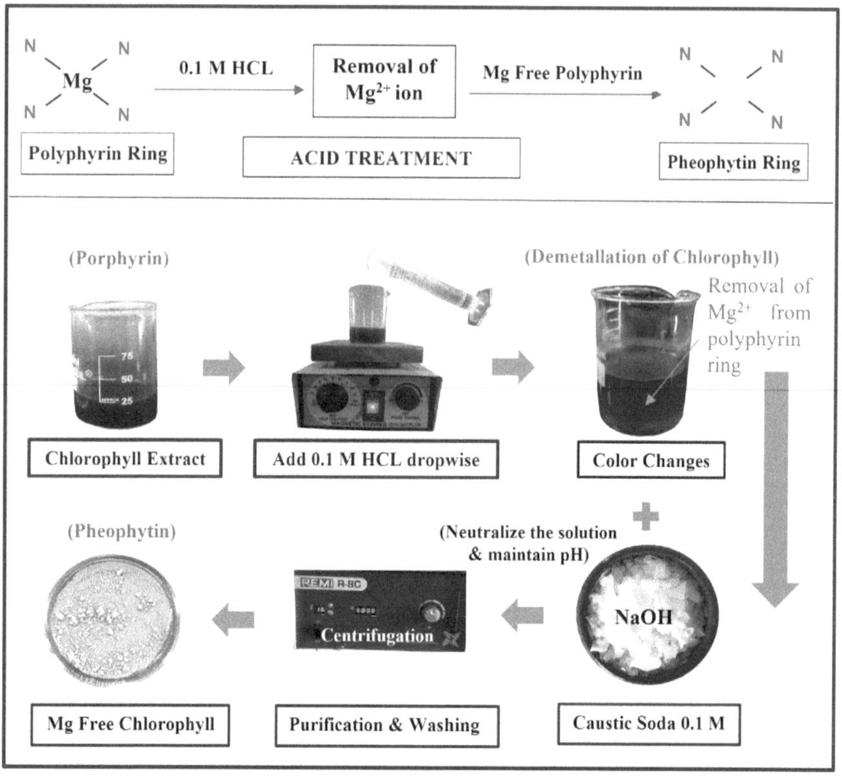

Figure 11.4 Demetalation process of eliminating Mg^{++} from chlorophyll.

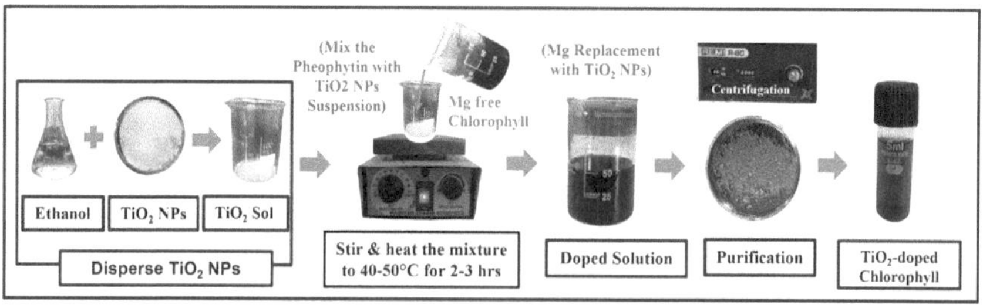

Figure 11.5 Incorporating TiO_2 NPs with pheophytin solution.

to record the absorption spectra in the wavelength range of 200–800 nm. The absorption spectra revealed distinct peaks corresponding to the light-harvesting capabilities of the hybrid materials.

11.3 RESULTS AND DISCUSSION

11.3.1 UV-Vis Absorption Spectrum of Chlorophyll

The UV-Vis absorption spectrum of chlorophyll extracted from Ficus religiosa leaves exhibits characteristic absorption peaks in the visible region, essential for understanding its light-harvesting capabilities. The spectrum displays two prominent absorption bands as shown in Figure 11.6. The strong absorption peak observed around 430 nm (blue region) arises from π-π^* transitions within the conjugated porphyrin ring system of the chlorophyll molecule (Borah & Bhuyan, 2017; Senge *et al.,* 2014). This peak corresponds to the Soret band, characteristic of porphyrin structures, and is associated with the strong absorption of higher-energy photons in the blue region of the spectrum. The other prominent absorption peak, located around 660 nm (red region), is attributed to n-π^* transitions within the porphyrin ring (Farajzadeh *et al.,* 2020; Lu & Kobayashi, 2016). This peak corresponds to the Q-band of chlorophyll and is associated with the absorption of lower-energy photons in the red region of the spectrum. The Q-band is particularly crucial for photosynthesis, as it captures light energy used to drive photosynthetic processes.

The broad absorption range of chlorophyll across the visible region (400–700 nm) highlights its ability to efficiently capture a wide range of solar radiation. This characteristic makes chlorophyll a promising candidate for organic photovoltaic applications. However, the relatively low absorption in the UV region suggests that modifications, such as doping with nanoparticles, may be necessary to enhance UV light absorption and improve the overall efficiency of chlorophyll-based solar cells.

11.3.2 UV-Vis Absorption Spectrum of TiO_2-Doped Chlorophyll

The UV-Visible absorption spectrum (as shown in Figure 11.7) of the TiO_2-doped chlorophyll reveals a broadened absorption range extending from 200 nm to 700 nm, encompassing both the ultraviolet and visible regions of the electromagnetic spectrum. The enhanced absorption in the UV region (200–400 nm) is attributed to the wide bandgap of the TiO_2 nanoparticles (approximately 3.2 eV), a characteristic that enables efficient harvesting of UV light. In the visible region (400–700 nm), the spectrum retains the characteristic absorption peaks of chlorophyll, specifically the Soret band

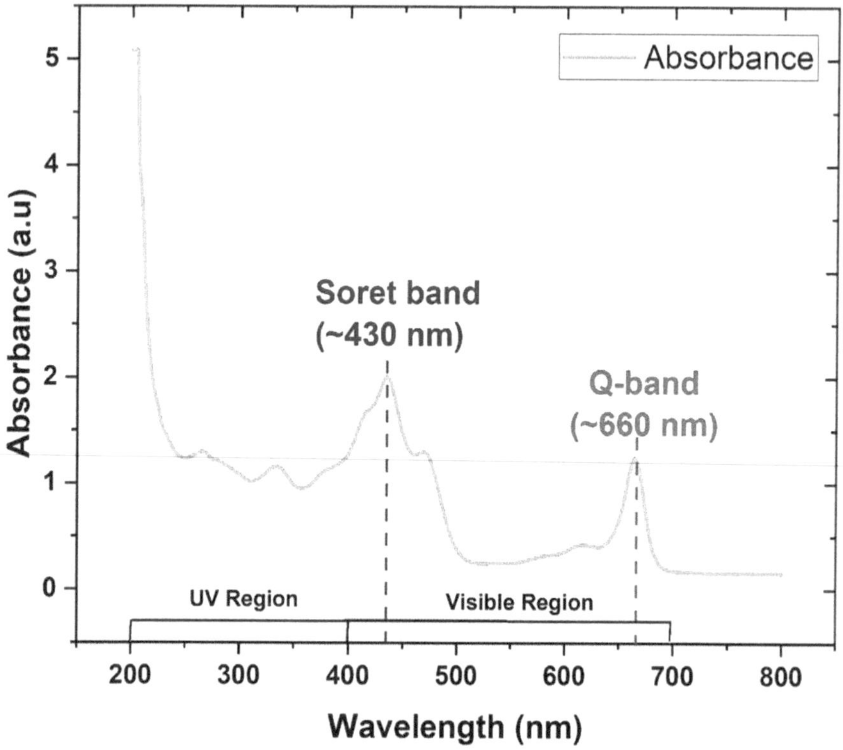

Figure 11.6 UV-Visible spectra of chlorophyll.

at ~430 nm and the Q-band at ~660 nm, indicating that the light-absorbing capabilities of chlorophyll are preserved after doping with TiO_2.

This broadened absorption spectrum, covering both UV and visible regions, makes the TiO_2-doped chlorophyll a highly promising material for solar energy applications, particularly in the fabrication of dye-sensitized solar cells (Yang, 2012; Brady, 2021). The enhanced absorption across a wider range of wavelengths suggests improved light harvesting, which can lead to increased photocurrent generation and potentially higher power conversion efficiency in photovoltaic devices. The TiO_2 nanoparticles act as a scaffold and electron acceptor, while the chlorophyll functions as the light-absorbing sensitizer. Upon light absorption, electrons are excited in chlorophyll and transferred to the TiO_2 conduction band, contributing to the photocurrent. This synergistic combination of TiO_2 and chlorophyll leverages the UV absorption of TiO_2 and the visible light absorption of chlorophyll, maximizing the utilization of the solar spectrum. Several studies have explored the use of TiO_2 in combination with various materials for enhanced solar cell performance (Guo & Ma, 2014; Quan *et al.*, 2005; Yue *et al.*, 2015).

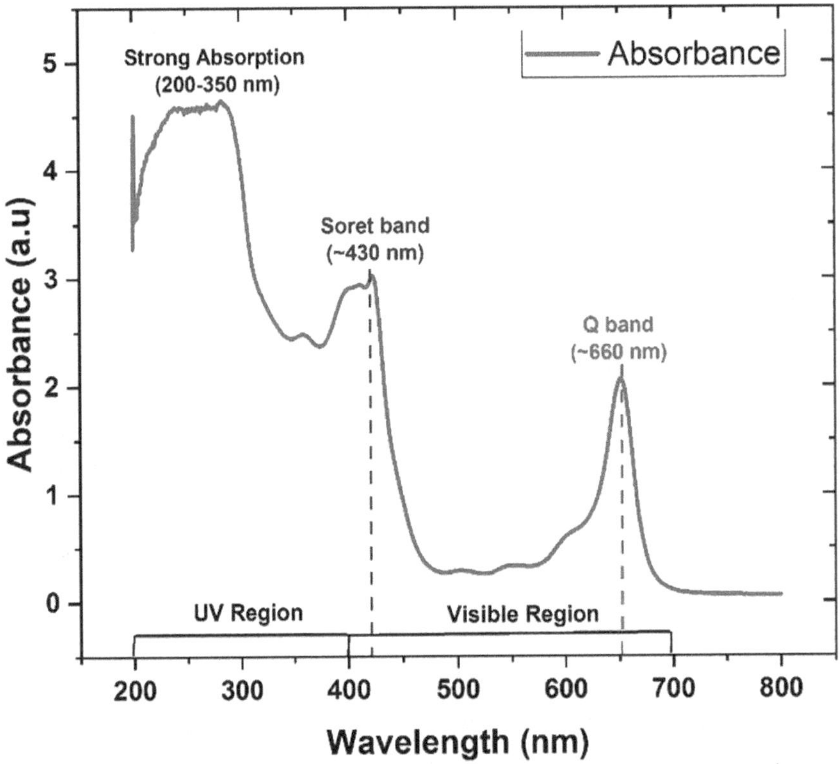

Figure 11.7 UV-Vis spectra of TiO_2-doped chlorophyll.

11.3.3 Comparative Analysis of Absorption Peak of UV-Vis Spectra

The UV-Visible spectra of chlorophyll and TiO_2-doped chlorophyll were analyzed to understand the impact of TiO_2 doping on the optical properties of chlorophyll. The absorption characteristics of both samples were compared, focusing on the UV and visible regions. The key differences observed in the absorption peaks, intensity, band broadening, and spectral shifts are shown in Table 11.1.

Table 11.1 summarizes the key optical changes observed, when chlorophyll is doped with TiO_2 nanoparticles, and how these changes can be leveraged for improved solar cell performance. As demonstrated in Figure 11.6, the pure chlorophyll exhibited weak absorption in the UV region (200 nm-350 nm), which means that a significant portion of the UV component of sunlight is not utilized for energy conversion in a chlorophyll-based solar cell. However, as observed in Figure 11.7, doping with TiO_2

Table 11.1 Comparative Assessment of Chlorophyll and Tio_2-Doped Chlorophyll

Parameters	*Chlorophyll*	*TiO_2-Doped Chlorophyll*	*Description*
UV Region Absorption	Weak absorption (200–350 nm)	Strong absorption (200–350 nm)	TiO_2 enhances UV absorption due to its intrinsic properties.
Visible Region Absorption	Strong peaks at ~430 nm (Soret band) and ~660 nm (Q band)	Modified peaks, possible shifts, or broadening	TiO_2 doping may alter the electronic structure, affecting visible absorption.
Band Broadening	Sharp and well-defined absorption bands	Broader absorption bands	Interaction between chlorophyll and TiO2 causes broadening.
Band Spectral Shifts	No significant shifts	Possible red or blue shifts	TiO_2 influences electronic transitions, leading to shifts in absorption peaks.
Additional Peaks	None	Possible new peaks in the UV region	TiO_2 introduces additional absorption features in the UV spectrum.

significantly enhances the UV absorption. TiO_2's intrinsic wide bandgap (~3.2eV) allows it to efficiently absorb UV light. This enhanced UV absorption is a significant advantage for solar cell applications, as it broadens the range of sunlight wavelengths that can be harvested, leading to increased photocurrent generation.

Likewise, chlorophyll shows strong absorption peaks in the visible region, specifically at ~430 nm and ~660 nm (Q band). These peaks correspond to specific electronic transition within the chlorophyll molecule. The TiO_2 doping can modify these peaks, potentially causing shifts or broadening. This is due to the interaction between chlorophyll and TiO_2 surface, which can alter the electronic environment of chlorophyll molecules; while these modifications can influence the visible light absorption, the chlorophyll still retains its ability to capture the light in this crucial region for photosynthesis and solar energy conversion. If we compare the band and band broadening, pure chlorophyll typically exhibits sharp, well-defined absorption bands, whereas the interaction between chlorophyll and TiO_2 leads to broader absorption bands. This broadening arises from the interaction of chlorophyll molecules with TiO_2 surface, creating a distribution of energy levels and broadening the absorption peaks. Broader absorption bands are generally beneficial for solar cells, as they allow for the capture of a wider range of wavelengths. The analysis in Figure 11.7 demonstrates TiO_2 doping can induce red and blue shifts in the absorption peaks. These shifts are due to the influence of TiO_2

on the electronic transition within the chlorophyll molecules. Optimizing these shifts can maximize light harvesting.

11.3.4 Bandgap Estimation of TiO_2-Doped Chlorophyll Using Tauc Plot

The optical bandgap of the TiO_2-doped chlorophyll was determined using the Tauc plot method. While TiO_2 is inherently an indirect bandgap material, the dominant optical behavior of the chlorophyll combined with quantum confinement effects from the TiO_2 nanoparticles results in the composite material exhibiting direct bandgap behavior. In an indirect bandgap material, the minimum of the conduction band and the maximum of the valence band occur at different momentum values. A transition between these bands requires a change in both energy and momentum. This typically involves interaction with a phonon (lattice vibration), making the transition less efficient than in direct bandgap materials. TiO_2 nanoparticles exhibit quantum confinement effects due to their small size. Quantum confinement occurs when the size of the material becomes comparable to the de Broglie wavelength of the electrons within it. This leads to a discretization of energy levels and can modify the bandgap energy, effectively making the material behave more like a direct bandgap material.

The Tauc plot method is used to determine the bandgap energy from the absorption spectrum. The absorption coefficient (α) is calculated using the Beer-Lambert law:

$$A = \alpha L c \tag{11.1}$$

where A is the absorbance, L is the path length, and c is the concentration. For a direct bandgap material, the Tauc relation is given by:

$$(\alpha h\vartheta)^2 = B(h\vartheta - E_g) \tag{11.2}$$

where h is Planck's constant, ϑ is the frequency of light, B is a constant, and E_g the bandgap energy.

By plotting $(\alpha h\vartheta)^2$ versus photon energy ($h\vartheta$) and extrapolating the linear region of the plot to the x-axis, the bandgap energy can be determined (Dolgonos *et al.*, 2016; Makuła *et al.*, 2018; Sangiorgi *et al.*, 2017; Zhong *et al.*, 2023).

The estimated bandgap of TiO_2-doped chlorophyll is 2.2 eV as shown in Figure 11.8, which is lower than the bandgap of pure TiO_2 (~3.2 eV). This reduction in bandgap is attributed to the incorporation of chlorophyll, which modified the electronic properties of the composite material. The lower bandgap enhances the material's ability to absorb visible light, making

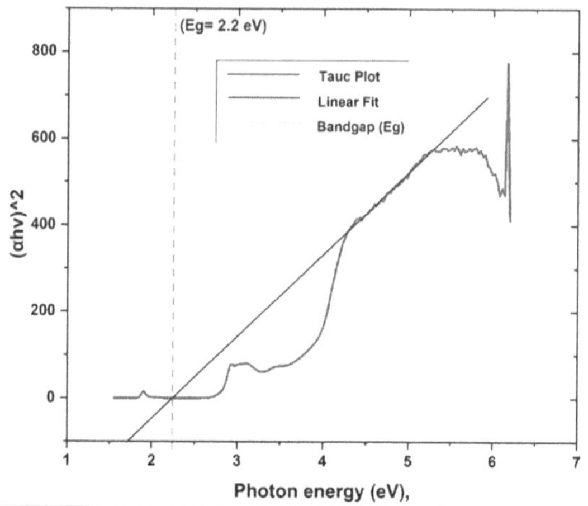

Figure 11.8 Bandgap estimation of TiO_2-doped chlorophyll.

it highly suitable for photovoltaic applications, particularly for visible light solar harvesting.

11.4 CONCLUSION AND FUTURE SCOPE

11.4.1 Conclusion

This study demonstrates the synthesis and characterization of a novel TiO_2-doped chlorophyll hybrid material with enhanced optical and electronic properties tailored for photovoltaic applications. The integration of TiO_2 nanoparticles with chlorophyll results in synergistic effects that improve light absorption across a broader range of the solar spectrum, encompassing both ultraviolet and visible regions. UV-Visible absorption spectroscopy revealed distinct absorption peaks corresponding to chlorophyll's characteristic Soret and Q bands in the visible region. The Soret band, typically located at ~430 nm, arises from electronic transitions within the chlorophyll molecule. The Q bands, typically at ~660 nm, are associated with other electronic transitions within the chlorophyll structure. The presence of TiO_2 nanoparticles introduces additional strong absorption in the UV region, effectively extending the material's light-harvesting capabilities beyond the visible spectrum. This broadened absorption is crucial for maximizing solar energy capture.

The estimated bandgap of 2.2 eV for the TiO_2-doped chlorophyll hybrid material is particularly significant for photovoltaic applications. This value falls within the optimal range for efficient solar energy harvesting in the visible spectrum. A lower bandgap compared to pure TiO_2 (~3.2 eV) indicates that the incorporation of chlorophyll modifies the electronic structure of the material, enabling it to absorb lower-energy photons in the visible region. This enhanced visible light absorption is key to improving the performance of photovoltaic devices. This research contributes to the development of sustainable and environmentally friendly materials for solar energy conversion. By combining natural components like chlorophyll with synthetic materials like TiO_2, this study offers a pathway toward cost-effective and high-performance photovoltaic devices. The hybrid material's ability to efficiently absorb a broad range of solar radiation positions it as a promising candidate for next-generation solar cells and other renewable energy technologies.

11.4.2 Future Scope

Future studies could focus on optimizing the concentration and distribution of TiO_2 nanoparticles within the chlorophyll matrix to further enhance the material's optical and electronic properties. Exploring different doping techniques and nanoparticle sizes could lead to improved performance.

The next step would involve fabricating and testing photovoltaic devices using the TiO_2-doped chlorophyll hybrid material. Investigating the material's performance in real-world solar cell configurations will provide valuable insights into its practical applicability.

Exploring the use of other nanomaterials, such as graphene, quantum dots, or other metal oxides, in combination with chlorophyll, could lead to the development of hybrid materials with even better properties.

REFERENCES

Adelekan, O. A., Ilugbusi, B. S., Adisa, O., Obi, O. C., Awonuga, K. F., Asuzu, O. F., & Ndubuisi, N. L. (2024). Energy transition policies: A global review of shifts towards renewable sources. Engineering Science & Technology Journal, 5(2), 272–287.

Albert, M. J. (2021). The climate crisis, renewable energy, and the changing landscape of global energy politics. Alternatives, 46(3), 89–98.

Amao, Y. (2008). Photoenergy conversion system based on the photosynthesis dyes conjugated nanoparticle. Current Nanoscience, 4(1), 45–52.

Arifin, Z., Soeparman, S., Widhiyanuriyawan, D., Suyitno, S., & Setyaji, A. T. (2017). Improving stability of chlorophyll as natural dye for dye-sensitized solar cells. Jurnal Teknologi, 80(1).

Arof, A. K., & Ping, T. L. (2017). *Chlorophyll as Photosensitizer in Dye-Sensitized Solar Cells*. InTech.

Benkerroum, N. (2019). Retrospective and prospective look at aflatoxin research and development from a practical standpoint. International Journal of Environmental Research and Public Health, 16(19), 3633.

Borah, K. D., & Bhuyan, J. (2017). Magnesium porphyrins with relevance to chlorophylls. Dalton Transactions, 46(20), 6497–6509.

Brady, N. G. (2021). Preserving the native environment of photosystem I within styrene maleic acid lipid particles (SMALPs) for applications in biohybrid solar devices. PhD diss., University of Tennessee, 2021. https://trace.tennessee.edu/utk_graddiss/6682.

Buscemi, G., Vona, D., Trotta, M., Milano, F., & Farinola, G. M. (2022). Chlorophylls as molecular semiconductors: Introduction and state of art. Advanced Materials Technologies, 7(2), 2100245.

Chandoliya, R., Sharma, S., Sharma, V., Joshi, R., & Sivanesan, I. (2024). Titanium dioxide nanoparticle: A comprehensive review on synthesis, applications and toxicity. Plants, 13(21), 2964.

Chen, X., Lou, Y., Dayal, S., Qiu, X., Krolicki, R., Burda, C., Zhao, C., & Becker, J. (2005). Doped semiconductor nanomaterials. Journal of Nanoscience and Nanotechnology, 5(9), 1408–1420.

Dey, S., Sreenivasulu, A., Veerendra, G. T. N., Rao, K. V., & Babu, P. A. (2022). Renewable energy present status and future potentials in India: An overview. Innovation and Green Development, 1(1), 100006.

Dhorkule, M., Lamrood, P., Ralegankar, S., Patole, S. P., Wagh, S. S., & Pathan, H. M. (2024). Unveiling the efficiency of dye-sensitized solar cells: A journey from synthetic to natural dyes. ES Food & Agroforestry, 16, 1086.

Dolgonos, A., Mason, T. O., & Poeppelmeier, K. R. (2016). Direct optical band gap measurement in polycrystalline semiconductors: A critical look at the Tauc method. Journal of Solid State Chemistry, 240, 43–48.

Duan, S., Zhou, Q., Li, A., Wang, X. F., Sasaki, S. I., & Tamiaki, H. (2020). Semi-synthetic chlorophyll derivatives toward solar energy applications. Solar RRL, 4(6), 2000162.

Farajzadeh, N., Akyüz, D., Koca, A., & Koçak, M. B. (2020). Electrochemical and spectroelectrochemical characterizations of new tetra-substituted metallophthalocyanines bearing 4-(trifluoromethoxy) phenoxy groups on non-peripheral positions. Monatshefte für Chemie—Chemical Monthly, 151, 1337–1348.

Fuziki, M. E. K., Tusset, Â. M., dos Santos, O. A. A., & Lenzi, G. G. (2023). Chlorophyll sensitization of TiO2: A mini-review. Reactions, 4(4), 766–778.

Guo, W., & Ma, T. (2014). Nanostructured nitrogen doping TiO2 nanomaterials for photoanodes of dye-sensitized solar cells. In *Low-Cost Nanomaterials: Toward Greener and More Efficient Energy Applications*, Z. Lin & J. Wang, Eds. (pp. 55–75). Springer, London.

Harris, D., et al. (2020). Exchange of the magnesium ion in chlorophyll by a transition metal. Chemistry, 1, 1–10.

Hassan, Q., Viktor, P., Al-Musawi, T. J., Ali, B. M., Algburi, S., Alzoubi, H. M., Al-Jiboory, A. K., Sameen, A. Z., Salman, H. M., & Jaszczur, M. (2024). The renewable energy role in the global energy transformations. Renewable Energy Focus, 48, 100545.

Hoerner, L. J. (2013). Photosynthetic solar cells using chlorophyll and the applications towards energy sustainability. USF St. Petersburg campus Honors Program Theses (Undergraduate). https://digitalcommons.usf.edu/honorstheses/136.

Iman, R. N., Younas, M., Harrabi, K., & Mekki, A. (2024). A comprehensive review on advancements and optimization strategies in dye-sensitized solar cells: Components, characterization, stability and efficiency enhancement. Inorganic Chemistry Communications, 112488.

Kalair, A., Abas, N., Saleem, M. S., Kalair, A. R., & Khan, N. (2021). Role of energy storage systems in energy transition from fossil fuels to renewables. Energy Storage, 3(1), e135.

Karafillidis, A. (2022). Chlorophylls. In *Comprehensive natural products III* (pp. 743–776). https://www.sciencedirect.com/referencework/9780081026915/comprehensive-natural-products-iii.

Koifman, O. I., Stuzhin, P. A., Travkin, V. V., & Pakhomov, G. L. (2021). Chlorophylls in thin-film photovoltaic cells, a critical review. RSC Advances, 11(25), 15131–15152.

Kumara, N. T. R. N., Lim, A., Lim, C. M., Petra, M. I., & Ekanayake, P. (2017). Recent progress and utilization of natural pigments in dye sensitized solar cells: A review. Renewable and Sustainable Energy Reviews, 78, 301–317.

Liang, J., Xiao, K., Wang, X., Hou, T., Zeng, C., Gao, X., Wang, B., & Zhong, C. (2024). Revisiting solar energy flow in nanomaterial-microorganism hybrid systems. Chemical Reviews, 124(15), 9081–9112.

Lu, H., & Kobayashi, N. (2016). Optically active porphyrin and phthalocyanine systems. Chemical Reviews, 116(10), 6184–6261.

Makuła, P., Pacia, M., & Macyk, W. (2018). How to correctly determine the band gap energy of modified semiconductor photocatalysts based on UV–Vis spectra. The Journal of Physical Chemistry Letters, 9(23), 6814–6817.

Mariotti, N., Bonomo, M., Fagiolari, L., Barbero, N., Gerbaldi, C., Bella, F., & Barolo, C. (2020). Recent advances in eco-friendly and cost-effective materials towards sustainable dye-sensitized solar cells. Green Chemistry, 22(21), 7168–7218.

Meng, D., Xue, J., Zhao, Y., Zhang, E., Zheng, R., & Yang, Y. (2022). Configurable organic charge carriers toward stable perovskite photovoltaics. Chemical Reviews, 122(18), 14954–14986.

Mony, C., Kaur, P., Rookes, J. E., Callahan, D. L., Eswaran, S. V., Yang, W., & Manna, P. K. (2022). Nanomaterials for enhancing photosynthesis: Interaction with plant photosystems and scope of nanobionics in agriculture. Environmental Science: Nano, 9(10), 3659–3683.

Musazade, E., Voloshin, R., Brady, N., Mondal, J., Atashova, S., Zharmukhamedov, S. K., Huseynova, I., Ramakrishna, S., Najafpour, M. M., Shen, J. R., & Bruce, B. D. (2018). Biohybrid solar cells: Fundamentals, progress, and challenges. Journal of Photochemistry and Photobiology C: Photochemistry Reviews, 35, 134–156.

Njeru, E. G., Njoroge, M. A., Kuria, K. P., & Mwangi, G. J. (2023). Leveraging on green energy harvesting via dye sensitized solar cells; their components, operations and graphene based electrodes: Renewable energy. International Journal of Emerging Trends in Science and Technology, 10, 8000–8031.

Omar, A., & Abdullah, H. (2014). Electron transport analysis in zinc oxide-based dye-sensitized solar cells: A review. Renewable and Sustainable Energy Reviews, 31, 149–157.

O'Regan, B., & Grätzel, M. (1991). A low-cost, high-efficiency solar cell based on dye-sensitized colloidal TiO2 films. Nature, 353(6346), 737–740.

Panagoda, L. P. S. S., Sandeepa, R. A. H. T., Perera, W. A. V. T., Sandunika, D. M. I., Siriwardhana, S. M. G. T., Alwis, M. K. S. D., & Dilka, S. H. S. (2023). Advancements in photovoltaic (Pv) technology for solar energy generation. Journal of Research Technology & Engineering, 4(30), 30–72.

Panwar, N. L., Kaushik, S. C., & Kothari, S. (2011). Role of renewable energy sources in environmental protection: A review. Renewable and Sustainable Energy Reviews, 15(3), 1513–1524.

Pimoli, D. S., Tomar, P., Kumar, A., Dwivedi, G., & Shukla, A. (2025). Experimental investigation of friction factor and heat transfer enhancement using circular perforated twisted tape inserts in heat exchangers. International Journal of Energy for a Clean Environment, 26(3). https://doi.org/10.1615/InterJEnerCleanEnv.2024053616.

Prakash, P., & Janarthanan, B. (2023). Review on the progress of light harvesting natural pigments as DSSC sensitizers with high potency. Inorganic Chemistry Communications, 152, 110638.

Putra, M. D., Darmawan, A., Wahdini, I., & Abasaeed, A. E. (2017). Extraction of chlorophyll from pandan leaves using ethanol and mass transfer study. Journal of the Serbian Chemical Society, 82, 921–931.

Quan, X., Yang, S., Ruan, X., & Zhao, H. (2005). Preparation of Titania nanotubes and their environmental applications as electrode. Environmental Science & Technology, 39(10), 3770–3775.

Rekioua, D., Mokrani, Z., Kakouche, K., Rekioua, T., Oubelaid, A., Logerais, P. O., . . . Ghoneim, S. S. 2023. Optimization and intelligent power management control for an autonomous hybrid wind turbine photovoltaic diesel generator with batteries. Scientific Reports, 13(1), 21830. http://dx.doi.org/10.1038/s41598-023-49067-4.

Roslan, N., Ya'acob, M. E., Radzi, M. A. M., Hashimoto, Y., Jamaludin, D., & Chen, G. (2018). Dye sensitized solar cell (DSSC) greenhouse shading: New insights for solar radiation manipulation. Renewable and Sustainable Energy Reviews, 92, 171–186.

Sangiorgi, N., Aversa, L., Tatti, R., Verucchi, R., & Sanson, A. (2017). Spectrophotometric method for optical bandgap and electronic transitions determination of semiconductor materials. Optical Materials, 64, 18–25.

Saxena, S., Dixit, F., Dalapathi, G. K., Krishnamurthy, S., & Kandasubramanian, B. (2022). Bioengineered solar harvesting systems for next generation applications. Solar Energy, 231, 857–879.

Selan, O. T. E., Hawu, H. N., Poy, L., Cunha, T. D., & Darmakusuma, D. (2021). The substitution of Mg^{2+} with $Mn^{2+/4+}$ metal ions in chlorophyll structure isolated from Gliricidia sepium leaves. Journal of Physics: Conference Series, 2017(1), 012002.

Senge, M. O., Ryan, A. A., Letchford, K. A., MacGowan, S. A., & Mielke, T. (2014). Chlorophylls, symmetry, chirality, and photosynthesis. Symmetry, 6(3), 781–843.

Senthilkumar, S., & Manimekalai, R. (2020). Growth and evaluation of linear optical, nonlinear optical, mechanical and thermal properties of a novel bio-organic chlorophyll-b of Ficus religiosa added ammonium dihydrogen phosphate optical single crystal. Optik, 212, 164679.

Shan, Z., Yi, Z., Fang, J., Fang, L., Lu, C., & Xu, Z. (2024). Advancing chlorophyll photostability: Dual physicochemical protection via Ce-doped hydrotalcite organic–inorganic hybrid pigments. ACS Applied Materials & Interfaces, 16(39), 52766–52779.

Singh, A., & Mukherjee, T. (2022). Application of carotenoids in sustainable energy and green electronics. Materials Advances, 3(3), 1341–1358.

Singh, A. R., Kumar, R. S., Bajaj, M., Khadse, C. B., & Zaitsev, I. (2024). Machine learning-based energy management and power forecasting in grid-connected microgrids with multiple distributed energy sources. Scientific Reports, 14(1), 19207.

Sugathan, V., John, E., & Sudhakar, K. (2015). Recent improvements in dye sensitized solar cells: A review. Renewable and Sustainable Energy Reviews, 52, 54–64.

Tamiaki, H., & Kichishima, S. (2025). Chlorophyll pigments and their synthetic analogs. Plant and Cell Physiology, 66(2), 153–167.

Tamrakar, E., Singla, R., Patel, R. N., Kumar, A., Tamrakar, R. K., & Butolia, S. (2024). Review on energy storage applications using new developments in solar photovoltaic materials and systems. Chemical Papers, 1–22.

Tiwari, S., Swaminathan, M., Eashwar S. S., Harender, & Singh, D. B. (2022). Performance enhancement of the photovoltaic system with different cooling methods. Environmental Science and Pollution Research, 29(30), 45107–45130. https://link.springer.com/article/10.1007/s11356-022-20330-x.

Xiong, W., Peng, Y., Ma, W., Xu, X., Zhao, Y., Wu, J., & Tang, R. (2023). Microalgae–material hybrid for enhanced photosynthetic energy conversion: a promising path towards carbon neutrality. National Science Review, 10(10), nwad200.

Yang, D., Wang, S., Liu, Y., Gao, J., Li, X., & Yuan, F. (2025). High-efficient naturally extracted chlorophyll modified CdS nanoparticles for pollutants degradation. Journal of the Australian Ceramic Society, 1–10. https://doi.org/10.1007/s41779-025-01183-2.

Yang, M. (2012). Band gap engineering and carrier transport in TiO_2 for solar energy harvesting (Doctoral Dissertation, University of Pittsburgh).

Yue, W., Liu, C., Qiu, Z., & Wang, M. (2015). TiO2/CuInS2-core/shell nanoarrays for polymer-based hybrid solar cells with aligned bulk heterojunctions. Solar Energy, 113, 358–368.

Zhan, W., Sun, L., & Han, X. (2019). Recent progress on engineering highly efficient porous semiconductor photocatalysts derived from metal–organic frameworks. Nano-Micro Letters, 11, 1–28.

Zhong, H., Pan, F., Yue, S., Qin, C., Hadjiev, V., Tian, F., Liu, X., Lin, F., Wang, Z., Ren, Z., & Bao, J. (2023). Idealizing Tauc plot for accurate bandgap determination of semiconductor with UV-Vis: A case study for cubic boron arsenide. arXiv preprint arXiv:2307.05371.

Chapter 12

Solar Thermal Collectors for Decentralized Cooking

A Review of Design and Implementation Strategies

Albar Khan, Akshat Garg, and Anoop Kumar Shukla

12.1 INTRODUCTION

In rural India, cooking is a key activity, consuming about 37% of the country's primary energy supply [1]. Approximately 70% of India's population lives in rural regions, where 75% of the energy for cooking is derived from non-commercial fuels such as firewood, which contributes to deforestation [2, 3]. However, India is rich in solar resources, with about 58% of its land receiving an average solar insolation of 5 kWh/m^2 per day annually [4], making solar cooking a promising alternative.

Recent advancements in solar thermal collector systems (SCS) have played a pivotal role in propelling the global agenda toward sustainable energy solutions. Over the past decade, innovations in this domain have significantly contributed to fulfilling several Sustainable Development Goals (SDGs), particularly SDG 7, which emphasizes universal access to clean, affordable, and modern energy. Moreover, solar thermal technologies support interconnected sustainability objectives, including SDG 13 (climate action), SDG 15 (life on land), and SDG 17 (partnerships for achieving goals).

Solar thermal technologies provide dynamic and adaptable energy alternatives, making them an integral component of decentralized energy frameworks. One crucial domain where their potential remains largely untapped is intermediate-temperature thermal applications (120–240°C), which encompass essential cooking techniques such as frying, roasting, and baking. These high-energy-demanding processes can be efficiently powered by solar thermal solutions, provided that thermal losses are mitigated to optimize heat retention. However, due to various limitations, solar cooking technologies have remained an area of ongoing exploration rather than widespread implementation. A key challenge that persists is the comprehensive performance evaluation of solar cookers designed for intermediate-temperature cooking, a factor often neglected in conventional assessments.

 DOI: 10.1201/9781003634737-12

The innovation and refinement of solar thermal collector (SC) designs are imperative in addressing energy scarcity, particularly in remote and decentralized locations where traditional power infrastructure is either absent or unreliable. The integration of multifunctional SC configurations introduces additional advantages, offering practical and sustainable energy solutions to communities operating independently of centralized grids.

This study presents the conceptualization, engineering, and development of an advanced solar thermal collector (SC) tailored for multifunctional applications, encompassing culinary preparation, dehydration of agricultural produce, and thermal fluid heating. The versatility of such a system establishes it as a transformative tool for augmenting decentralized clean energy access.

Despite its multifunctional capacity, this study primarily underscores the SC's utility in culinary applications for the following compelling reasons:

- Cooking is an indispensable daily activity with a substantial energy footprint. Traditional cooking fuels such as wood, charcoal, and liquefied petroleum gas contribute to environmental degradation and economic burdens. Solar-powered cooking solutions can alleviate these pressures while ensuring sustainability.
- In isolated and stand-alone settings, residential layouts typically feature expansive sunlit spaces, such as courtyards, terraces, or backyards, which provide optimal conditions for SC installation. The convenience of direct placement enhances accessibility and ease of use.
- These open-air environments allow maximum solar irradiation capture, fostering diverse energy applications beyond cooking, such as food drying and water heating, without stringent spatial constraints.
- Cutting-edge SC innovations, including hybridized systems capable of sustaining intermediate temperature ranges, exhibit extraordinary potential as multifunctional utilities adaptable for various domestic and commercial applications.
- While previous study endeavors have explored solar cookers' thermal efficiencies, assessments have primarily revolved around broad temperature classifications (low, medium, and high) without delving into intricate performance parameters. A more rigorous evaluation, incorporating energy and exergy efficiency analyses, is essential to bridge the existing study gap and enhance SC's practical efficacy.

By leveraging meticulous engineering principles, material optimization, and novel hybridization strategies, this work endeavors to set a precedent for next-generation SC systems, transforming decentralized energy landscapes while fostering economic, environmental, and societal resilience.

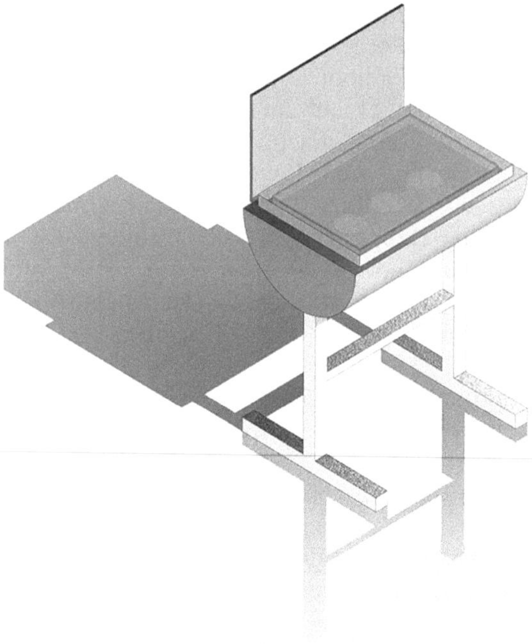

Figure 12.1 3D model of solar thermal collector (half-cylinder [Tambor]).

12.2 DESIGN, CONSTRUCTION, AND WORKING OF SOLAR THERMAL COLLECTOR

Figure 12.1 illustrates the 3D model of the half-cylinder (Tambor) solar thermal collector (referred to as HTSC), initially designed by Oscar Núñez Martínez [5] and later modified by our team. This collector will be constructed using a horizontally cut metallic cylinder with a diameter of 48 cm.

The thickness of the metallic structure will range between 1.2 mm and 1.5 mm. The overall dimensions of the cylinder are 725 mm in length and 480 mm in width. A rectangular absorber plate, measuring 0.89 m^2 (580 mm in length and 170 mm in width) and with a thickness between 1.2 mm and 1.5 mm, will be constructed from black paving material and installed inside the cylinder with the help of a metallic support stand.

The interior of the cylinder will be equipped with a highly polished stainless-steel reflector, matching the size of the cylinder. Between the stainless-steel reflector and the cylinder's surface, 40–50 mm thick glass wool

Table 12.1 Material Used, Specs, and Other Props of the HTSC

Metallic Half-Cylinder (Tambor-HTSC)	
Material used	Galvanized iron
HTSC diameter	Ø 480 mm
HTSC effective aperture area	Ø 585 mm × 375 mm
HTSC thickness	1.2 mm to 1.5 mm
HTSC length	650 mm
Glazed Double Glass Cover	
Glass cover size	585 mm × 375 mm
Each glass thickness	6 mm
Glass gaps in between them	7 mm
Glass transmittance	0.85–0.95
Stainless-Steel Reflectors (Booster & Internal)	
Booster reflector effective size	585 mm × 375 mm
Booster & internal reflector thickness	1.2 mm to 1.5 mm
Diameter of internal reflector	375 mm
Internal reflector length	585 mm
Absorber Plate	
Absorptivity of absorber plate painted with black	0.85–0.90
Size of absorber plate	525 mm × 170 mm
Absorber plate material	Aluminum
Absorber plate thickness	1 mm
Insulating Material	
Material	Glass wool
Insulation thickness	40–50 mm
Thermal conductivity at 100°C	0.07–0.044 W/mK

insulation will be press-fitted to minimize heat loss and improve heat retention. This internal reflector will enhance the solar radiation directed toward the absorber plate and other components, such as cooking pots or food being dried.

The top of the half-cylinder will be covered with a double-glass cover, each glass measuring 3–5 mm in thickness and sized at 585 mm × 375 mm, totaling an area of 0.219 m^2. There will be a 7 mm gap between the two glass layers, which will be held apart by a metallic support. The metallic support and glass will be tightly sealed using high-temperature silicone to prevent heat loss.

Additionally, a rectangular booster reflector/mirror, made of highly polished stainless steel, will be attached along the length of the half-cylinder using a long hinge. The booster mirror will have an effective aperture area of 585 mm × 375 mm (0.219 m^2) and a total aperture area of 0.34 m^2. The tilt

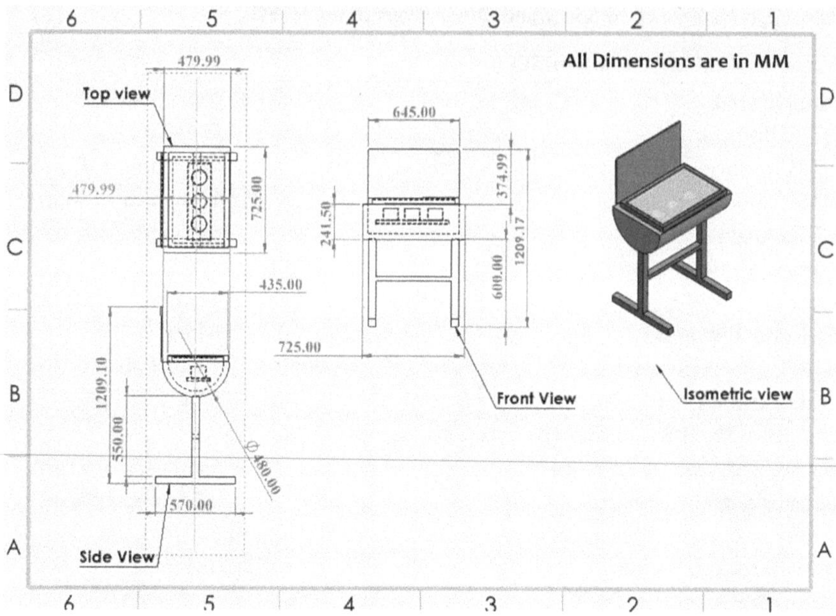

Figure 12.2 2D Sketch of solar thermal collector.

of the mirror can be adjusted between 0° and 120°, allowing seasonal adjustments to optimize solar capture. Table 12.1 outlines the materials, properties, and specifications of the various components of the HTSC. Figure 12.2 depicts the 2D sketch of the HTSC, showing the front, top, and side views. Figure 12.3 presents the bill of materials (BOM) for the HTSC.

12.3 THEORETICAL BACKGROUND FOR TESTING THE TAMBOR-SHAPED SOLAR COLLECTOR

The effectiveness of a particular cooker is assessed based on its thermal performance and specific objective criteria. Several thermal performance indicators are available to evaluate intra-cooker performance, including the figure of merit (F1) [6], efficiency, cooking power, and the heat loss factor ($F'U_L$) [7], along with the optical efficiency-factor ($F'\eta_O$) [7].

The Hottel-Whillier-Bliss (HWB) equation for solar collectors shows that the optical efficiency factor ($F'\eta_O$) and the heat-loss factor ($F'U_L$) are key factors in determining the thermal efficiency of a box cooker (BC).

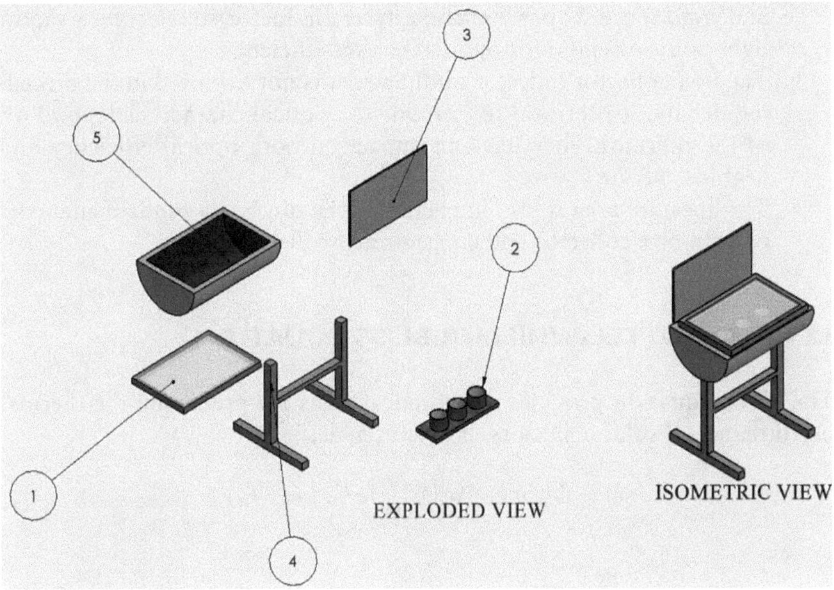

Figure 12.3 BOM of the HTSC.

Item No.	*Part Name*	*Description*	*QTY.*
1	Glass lid	Double Glazed Glass Cover [585 × 375 mm]	1
2	Pots with absorber plate	Aluminum pots for cooking with absorber plate painted with black [525 × 170 mm]	1
3	Reflector	High Gloss stainless steel reflector [585 mm × 375 mm] 1.2 mm to 1.5 mm thick	1
4	Stand	Metallic stand made of Iron	1
5	Tambor	Half cut Metallic cylinder made of galvanized iron [Ø 200 mm]	1

- The optical efficiency factor ($F'\eta_O$) reflects the cooker's maximum performance level and is influenced by the optical properties of the glass cover, absorber, and cooking pot, as well as tracking accuracy.
- The heat-loss factor ($F'U_L$) depends on design features and operating conditions, increasing with the temperature of the cooking pot's water

and wind speed. Vapor loss also affects this factor, so selecting a vapor-tight pot is essential for optimal cooker efficiency.

- The box collector reflector's influence was not explored in the preceding debate. It is critical to examine the optical characteristics and tilt of the reflector. They have an impact on both optical efficiency and heat loss factor.
- The aperture area is also impacted by the tilt. It is a fundamental part of both box collector and concentrated collector.

12.4 THE HOTTEL-WHILLIER-BLISS EQUATION

The HWB equation provides a theoretical basis for predicting the thermal performance of solar collectors. It is written as:

$$Q_u = F_R A C\left[(\tau\alpha)_e I - U_L\left(T_{fi} - T_a\right)\right]$$

where:

Q_u = Useful heat gain by the fluid (W)
F_R = Heat removal factor
A_C = Collector area (m^2)
$(\tau\alpha)_e$ = Effective transmittance-absorptance product
I = Incident solar radiation (W/m^2)
U_L = Overall heat loss coefficient (W/m^2K)
T_{fi} = Inlet fluid temperature (°C)
T_a = Ambient temperature (°C)

Equation 12.1 represents the rate of useful heat gained by the test load per unit of aperture area (Ap) of the thermal solar collector and per unit of time during heating experiments [8, 9];

$$Q'' = \frac{\left[\left(m_f . C_{pf}\right).\left(T_{f2} - T_{f1}\right)\right]}{A_p . \Delta t} \tag{12.1}$$

where:

- Q'' : Rate of useful heat gain per unit area of the aperture (W/m^2).
- mf : Mass of the fluid used as the test load (kg).
- C_{pf} : Specific heat capacity of the test fluid (J/(kg·°C)).
- $T_{f2} - T_{f1}$: Change in the fluid's temperature between the initial (T_{f1}) and final (T_{f2}) states (°C).
- A_P : Aperture area of the solar cooker (m^2).
- Δt : Time interval over which the temperature change occurs (seconds).

Any fluid that could be utilized as a test load for thermal solar collector testing is denoted by the letter (f). Water's lower boiling point and consequent problems with fluid dynamic qualities make it unsuitable for use as a test fluid.

12.5 COOKER OPTICAL THERMAL RATIO

The cooker optical thermal ratio (COTR) is a parameter used in the analysis of solar cookers. Also, COR has been used extensively in the past literature [10, 11–17].

It essentially measures the efficiency of a solar cooker by relating the optical performance of the cooker to its thermal performance.

Optical Performance: This refers to how well the cooker collects and directs sunlight onto the cooking surface. It depends on factors like the reflectivity of the mirrors, the transmissivity of the glass, and the absorptivity of the cooking pot or the absorber plate.

Thermal Performance: This relates to the heat retention and heat transfer characteristics of the solar cooker. It involves aspects such as insulation, heat losses (conduction, convection, and radiation), and the thermal capacity of the materials used.

Equation 12.2 calculates the COR for the suggested design of the solar collector.

$$COR = \frac{F'\eta O}{\frac{F'U_l}{C}} = \frac{\eta OC}{U_l} \tag{12.2}$$

where η_O is the optical efficiency, C is the concentration ratio, and U_L is the factor of heat loss.

12.6 GEOMETRIC AND EFFECTIVE CONCENTRATION RATIO

The geometrical concentration ratio (C_{GCR}) of the thermal solar collector is a ratio of the inclined aperture-area (A_{tilt}) to the glaze heat-loss area (A_g) (See Figure 12.3) and expressed using Equation 12.3:

$$C_{GCR} = \frac{A_{tilt}}{A_g} \tag{12.3}$$

The effective concentration ratio (C_{EF}) of HTSC is the ratio of cooker optical thermal ratios determined with and without booster reflector and is represented using Equation 12.4:

$$C_{EF} = \frac{COR}{COR_{NBR}} \tag{12.4}$$

CEF allows the assessment of the influence of added solar flux reflected by booster reflector and its function in the performance improvement of the thermal solar collector. In Equation 12.4, COR is cooker opto-thermal ratio and NBR means no booster reflector. In the present case, the value of Atilt varies between $0.32m^2$ and $0.39m^2$ with mean value of ~$0.35m^2$. The dimensions for this calculation are AB = 645 mm, BC = 605 mm, DE = 475 mm, AE = 375 mm, Ag = 0.22 mm^2. So, the concentration ratio for the thermal solar collector comes out between 1.45 and 1.77 with the mean value of 1.61.

12.7 OPTIMUM TILT ANGLE OF THE BOOSTER MIRRORS

The optimum tilt angle [(18)] of the booster mirrors is essential for maximizing the reflected solar radiation onto the absorber plate of a solar cooker. In the context of this paper, the booster mirrors' tilt angle is adjusted for both horizontally placed cookers and inclined cookers to ensure that the maximum amount of reflected solar energy reaches the absorber plate, especially during different seasons.

Figure 12.4 Thermal solar collector's inclined aperture area (ABCD) and glazed heat loss area (DCFE).

12.7.1 Horizontally Placed Cooker

For a horizontally placed cooker, the **optimum tilt angle** (λ) of the booster mirror is determined based on the following geometry:

Equation 12.5 for Tilt Angle:

$$\lambda + \theta z + i = 90^{\circ} \quad (12.5)$$

where:

- λ = Optimum angle of the booster mirror
- θz = Sun's zenith angle
- i = Incident angle of the sun's rays

This equation ensures that the reflected rays from the booster mirror hit the absorber plate at the correct angle.

Calculation of λ:

$$\text{From the equation}[6]: \lambda = \frac{90^{\circ} - 2\theta_z}{3} \quad (12.6)$$

This formula gives the optimal tilt angle for the booster mirror based on the zenith angle of the sun.

12.8 FINDING OUT THE ZENITH ANGLE

The **zenith angle** (θ_Z) is the angle between the sun and a vertical line from a specific location on the Earth's surface. It is an important parameter for solar energy calculations because it tells us how high the sun is in the sky. When the zenith angle is 0°, the sun is directly overhead.

12.9 FORMULA TO CALCULATE THE ZENITH ANGLE

The zenith angle is calculated using the following formula:

$$\theta_z = cos^{-1}\left(sin\phi \times sin\delta + cos\phi \times cos\delta \times cos\omega\right) \quad (12.7)$$

where:

- θ_z = Zenith angle
- ϕ = Latitude of the location

- δ = Declination angle of the sun
- ω = Hour angle

The Parameters:

1. **Latitude (ϕ):**

   The latitude of the location where the solar cooker is placed. This is a constant for the location and can be found using maps or GPS.

2. **Declination-Angle (δ):**

   The declination angle depends on the day of the year and is calculated using the following formulae:

$$\frac{\delta = 23.45^{o} \,.\, \sin\left(360^{o}.\,(284+n)\right)}{365}$$

   where n is the number of days given that the beginning of the year (January 1st = day 1).

3. **Hour Angle (ω):**

   The hour angle (ω) indicates how far the sun is from the local meridian (solar noon). It is calculated as:

$$\omega = 15^{o}.\left(t_{solar} - 12\right)$$

   where *tsolar* is the solar time (hours). For example, at solar noon, the hour angle is 0°, and at 1 hour before solar noon, the hour angle is –15°, while 1 hour after solar noon, it is +15°.

An Example:

Given Data:

- Latitude = 28.541235°
- Local Time = 10:00:00
- Solar Noon = 12:15:36
- Declination Angle = –3.31° (for September 16, 2024)
- Time Zone = Asia/Kolkata (UTC +5:30)

12.9.1 Calculate the Hour Angle at 10:00 AM

The hour angle (ω) is the angular translation of the sun from the restricted meridian due to the Earth's rotation. The hour angle is calculated using the difference between the local time and solar noon. Each hour corresponds to 15°.

- Local Time = 10:00:00
- Solar Noon = 12:15:36

We first find the time difference in hours between 10:00:00 AM and 12:15:36 PM:

$$Time\,Difference = (12 + 15/60 + 36/3600) - 10 = 2.26\ hours$$

Since this is before solar noon, the hour angle is negative. The hour angle is calculated as:

$$\omega = 15° \times \text{Time Difference} = 15° \times (-2.26) = -33.9°$$

12.9.2 Convert Latitude and Declination to Radians

We need to convert both the latitude and declination angle to radians for the trigonometric calculations.

$$Latitude = 28.541^{o} = \frac{28.541.\pi}{180} = 0.498\,radians$$

$$Declination = -3.31^{o} = \frac{-3.31.\pi}{180} = -.0578\,radians$$

12.9.3 Zenith Angle Formula

The formula for the **zenith angle** (θ_Z) is:

$$cos(\theta_z) = sin(Latitude) \cdot sin(Declination) + cos(Latitude) \cdot cos(Declination) \cdot cos(Hour\,Angle)$$

$$cos(\theta_z) = sin(0.4982) \cdot sin(-0.0578) + cos(0.4982) \cdot cos(-0.0578) \cdot cos(-33.9^{o})$$

12.9.4 Calculate Each Term

$$sin(Latitude) = sin(0.4982) = 0.4783$$

$$sin(Declination) = sin(-0.0578) = -0.0577$$

$$cos(Latitude) = cos(0.4982) = 0.8782$$

$$cos(Declination) = cos(-0.0578) = 0.9983$$

$$cos(Hour\ Angle) = cos(-33.9^o) = 0.8319$$

Now, calculate the cosine of the zenith angle:

$$\cos(\theta_z) = (0.4783 \times -0.0577) + (0.8782 \times 0.9983 \times 0.8319)$$

$$\cos(\theta_z) = -0.0276 + 0.7295 = 0.7019$$

12.9.5 Find the Zenith Angle

The zenith angle is:

$$\theta_z = \cos^{-1}(0.7019) = 45.51^\circ$$

Therefore,

At 10:00 AM local time on September 16, 2024, at latitude 28.541235°, the zenith angle is approximately 45.51°.

The formula to calculate λ (the optimum angle of the booster mirror) is:

$$\lambda = \frac{90^\circ - 2\theta_z}{3}$$

Example for 8:00 AM: Zenith Angle = 69.0076°

$$\lambda = \frac{[90^\circ - (2 X 69.0076)]}{3} = \frac{90^\circ - 138.0512}{3} = -\frac{48}{3} = -16^\circ$$

Here's a C++ **code** that calculates the **zenith angle** at different hours of the day for a given location, specifically at **latitude** 28.541235° N (Amity University Noida).

```
#include <iostream>
#include <cmath>
#include <ctime>
#define PI 3.14159265358979323846
// Function to calculate the hour angle
double hour_angle(int local_hour, int local_minute, int
local_second, int solar_noon_hour, int
solar_noon_minute, int solar_noon_second)
{// Convert the time difference to fractional hours
```

```
 double time_difference = (local_hour + local_minute / 60.0 +
 local_second / 3600.0)-
                  (solar_noon_hour + solar_noon_minute / 60.0
                  + solar_noon_second /3600.0);
   return 15.0 * time_difference; // Each hour corresponds to
   15 degrees
}
// Function to calculate the zenith angle
double zenith_angle(double latitude, double declination, dou-
ble hour_angle)
{ double latitude_rad = latitude * PI / 180.0;
  double declination_rad = declination * PI / 180.0;
  double hour_angle_rad = hour_angle * PI / 180.0;
// Zenith angle formula
double cos_zenith = sin(latitude_rad) *
sin(declination_rad) +
            cos(latitude_rad) * cos(declination_rad) *
cos(hour_angle_rad);
// Calculate the zenith angle in degrees
double zenith = acos(cos_zenith) * 180.0 / PI;
return zenith;
}
int main()
{ // Latitude and declination angle for 16th September
2024
  double latitude = 28.541235;
  double declination = -3.31; // Declination angle of the sun
// Solar noon time (12:15:36) in hours, minutes, sec-
onds
  int solar_noon_hour = 12;
  int solar_noon_minute = 15;
  int solar_noon_second = 36;
// Loop through different times of the day (e.g., from
8 AM to 4 PM)
  for (int hour = 8; hour <= 16; ++hour) {
    int minute = 0; // Assuming exact hours for
simplicity
    int second = 0;
// Calculate the hour angle for the given time
  double h_angle = hour_angle(hour, minute, second,
solar_noon_hour, solar_noon_minute, solar_noon_second);
// Calculate the zenith angle
```

```
  double z_angle = zenith_angle(latitude, declination, h_angle);
// Output the result
   std::cout << "At " << hour << ":00, Hour Angle = " <<
h_angle << " degrees, Zenith Angle = " << z_angle << "
degrees\n";
  }
  return 0;
}
```

Output:

At 8:00, Hour Angle = –63.9 degrees, Zenith Angle = 69.0076 degrees
At 9:00, Hour Angle = –48.9 degrees, Zenith Angle = 56.7059 degrees
At 10:00, Hour Angle = –33.9 degrees, Zenith Angle = 45.5457 degrees
At 11:00, Hour Angle = –18.9 degrees, Zenith Angle = 36.6653 degrees
At 12:00, Hour Angle = –3.9 degrees, Zenith Angle = 32.0711 degrees
At 13:00, Hour Angle = 11.1 degrees, Zenith Angle = 33.5903 degrees
At 14:00, Hour Angle = 26.1 degrees, Zenith Angle = 40.5366 degrees
At 15:00, Hour Angle = 41.1 degrees, Zenith Angle = 50.7064 degrees
At 16:00, Hour Angle = 56.1 degrees, Zenith Angle = 62.5122 degrees

Performance of solar collector under no-load condition and determination of first figure of merit F1:

- The **first test** is a stagnation test, which measures the maximum temperature that a cooker can achieve under no-load conditions.
- This test determines the **figure of merit first for thermal performance:** the ratio of the cooker's **optical efficiency** to its **heat-loss factor.**
- A high optical efficiency and a low heat loss factor are desirable.
- The cooker is placed in the sun, and the temperature is monitored until it reaches a stable point, called the **stagnation temperature.**

12.10 ENERGY BALANCE AT STAGNATION

An energy balance equation is presented for the cooker at stagnation, where there is no useful heat gain. The equation is:

$$\eta_o H_s = U_L\left(T_{ps} - T_a\right) \tag{12.8}$$

Table 12.2 Table for September 16, 2024

Time	Zenith Angle (θ_z)°	Optimum Angle (λ)°
8:00	69.0076	-16.0051
9:00	56.7059	−7.1372
10:00	45.5457	−0.3648
11:00	36.6653	5.8898
12:00	32.0711	8.6193
13:00	33.5903	7.22731
14:00	40.5366	3.3089
15:00	50.7064	−3.1375
16:00	62.5122	−10.3415

where:

- η_o is the optical efficiency,
- H_s is the **insolation** (solar radiation) on a horizontal surface,
- P_s is the **plate stagnation temperature,**
- U_L is the **overall heat loss coefficient,**
- T_a is the **ambient temperature** when stagnation temperature is reached.

An estimate made in this study is that the heat-loss factor is supposed to be constant.

12.11 FIGURE OF MERIT FOR HTSC

Mullick [(**6**)] proposed no-load-test-based performing assessment method and constraint figure of merit first (F_1) for the box-type solar collectors. Therefore, it is used in the solar cooker design here to evaluate performance under no load condition.

F_1 [(**6**)] is expressed as

$$F_1 = \frac{T_{ps} - T_a}{H_s} \tag{12.9}$$

This value indicates the cooker's ability to achieve high temperatures with minimal heat loss.

12.12 MINIMUM PERFORMANCE CRITERIA

A minimum value of F_1 can be set to ensure the cooker's thermal performance is sufficient. For example, if $F_1 \geq 0.12$, and under certain conditions (e.g., $H_s = 800\,W/m^2$, $T_a = 15°C$), the stagnation temperature must be at least 111°C. This ensures the cooker can achieve temperatures high enough for boiling-type cooking.

$$\frac{T_{ps} - T_a}{H_s} \geq 0.12 \ OR \ T_{ps} \geq 111^{\circ}C$$

This grading is according to the standards put by the Bureau of Indian Standards (BIS).

IS Number: IS 13429 (Part 1): 2020
Product: Solar Cooker—Box Type
Features: This standard (Part 1) specifies the requirements of box-type solar cookers for cooking. The solar cookers covered in this standard shall be of two grades depending upon their thermal performance (F_1). (a) Grade A: F_1 shall not be less than 0.12, and (b) Grade B: F_1 shall not be less than 0.11. This standard also covers the assembly, workmanship, components, tests, instruction manual of cookers, etc.

12.13 CONCLUSION AND FUTURE SCOPE

The conclusion is divided into four subsections for a detailed breakdown of HTSC:

- HTSC's no load condition performance and approximation of F_1; hence, the criteria for the no load test and the equation to be used are found. After the fabrication, the reading of the test will be used to find the stagnation temperature and the HTSC will be graded according to BIS standards.
- The optimal tilt of the booster reflector of the HTSC, and it was found earlier in the report.
- COR-based performance evaluation of the HTSC.
- The calculation of the effective concentration ratio will be completed using formulae derived after fabrication and analysis of readings.

12.14 NOMENCLATURE AND ABBREVIATIONS

Nomenclature	
C_{EF}	Concentration ratio
COR	Cooker opto-thermal ratio [(m^2.°C)/W]
$F'_{\eta 0}$	Optical efficiency factor (–)
F'_{UL}	Heat loss factor [W/(m^2.°C)]
F_I	First figure of merit
A_p	Aperture area
Abbreviations	
SC	Solar thermal collector
TPP	Thermal performance parameter
GCSC	Generalized cost of solar-cooked meals
HTSC	Half-cylinder (Tambor) solar thermal collector

REFERENCES

[1] Prasanna, U. R., & Umanand, L. (2011). Optimization and design of energy transport system for solar cooking application. *Applied Energy*, *88*(1), 242–251.

[2] Ramachandra, T. V. (2010). Mapping of fuelwood trees using geoinformatics. *Renewable and Sustainable Energy Reviews*, *14*(2), 642–654.

[3] Balachandra, P. (2011). Modern energy access to all in rural India: An integrated implementation strategy. *Energy Policy*, *39*(12), 7803–7814.

[4] Ramachandra, T. V., Jain, R., & Krishnadas, G. (2011). Hotspots of solar potential in India. *Renewable and Sustainable Energy Reviews*, *15*(6), 3178–3186.

[5] Sagade, A. A., Gálvez, E., Saxena, A., Martínez, O. N., & Belgasim, B. (2024). Enhancing clean energy access in decentralized communities through novel design of solar thermal collector: Experimental validation through intermediate temperature cooking. *Sustainable Energy Technologies and Assessments*, *71*, 103988.

[6] Lahkar, P. J., & Samdarshi, S. K. (2010). A review of the thermal performance parameters of box type solar cookers and identification of their correlations. *Renewable and Sustainable Energy Reviews*, *14*(6), 1615–1621.

[7] Sagade, A. A., Samdarshi, S. K., & Panja, P. S. (2018). Experimental determination of effective concentration ratio for solar box cookers using thermal tests. *Solar Energy*, *159*, 984–991.

[8] Lahkar, P. J., Bhamu, R. K., & Samdarshi, S. K. (2012). Enabling inter-cooker thermal performance comparison based on cooker opto-thermal ratio (COR). *Applied Energy*, *99*, 491–495.

[9] Sagade, A. A., Samdarshi, S. K., & Panja, P. S. (2018). Enabling rating of intermediate temperature solar cookers using different working fluids as test loads and its validation through a design change. *Solar Energy*, *171*, 354–365.

[10] Sagade, A. A., Mawire, A., Palma-Behnke, R., & Sagade, N. A. (2023). Low-cost solar concentrating collector-receiver system as an effective enabler for clean cooking and heating in urban areas. *Solar Energy*, *263*, 111813.

[11] Coccia, G., Di Nicola, G., Pierantozzi, M., Tomassetti, S., & Aquilanti, A. (2017). Design, manufacturing, and test of a high concentration ratio solar box cooker with multiple reflectors. *Solar Energy*, *155*, 781–792.

[12] Apaolaza-Pagoaga, X., Sagade, A. A., Ruivo, C. R., & Carrillo-Andrés, A. (2021). Performance of solar funnel cookers using intermediate temperature test load under low sun elevation. *Solar Energy*, *225*, 978–1000.

[13] Sagade, A. A., Apaolaza-Pagoaga, X., Ruivo, C. R., & Carrillo-Andrés, A. (2022). Concentrating solar cookers in urban areas: Establishing usefulness through realistic intermediate temperature rating and grading. *Solar Energy*, *241*, 157–166.

[14] Sagade, N. A., Sagade, A. A., Tawfik, M. A., & Saxena, A. (2023). Ensuring self-sustainability in decentralized communities through solar cookers combined with food dryers. *Solar Energy*, *260*, 83–93.

[15] Khallaf, A. M., Tawfik, M. A., El-Sebaii, A. A., & Sagade, A. A. (2020). Mathematical modeling and experimental validation of the thermal performance of a novel design solar cooker. *Solar Energy*, *207*, 40–50.

[16] Tawfik, M. A., Sagade, A. A., Palma-Behnke, R., Hanan, M., & Abd Allah, W. E. (2021). Solar cooker with tracking-type bottom reflector: An experimental thermal performance evaluation of a new design. *Solar Energy*, *220*, 295–315.

[17] Tawfik, M. A., Sagade, A. A., Palma-Behnke, R., Abd Allah, W. E., & Hanan, M. (2022). Performance evaluation of solar cooker with tracking type bottom reflector retrofitted with a novel design of thermal storage incorporated absorber plate. *Journal of Energy Storage*, *51*, 104432.

[18] Sethi, V. P., Pal, D. S., & Sumathy, K. (2014). Performance evaluation and solar radiation capture of optimally inclined box type solar cooker with parallelepiped cooking vessel design. *Energy Conversion and Management*, *81*, 231–241.

Chapter 13

Review of Advances in Solar Dryer Design and Technologies

Neetu Singh, Dr. V. N. Bartaria,
Dr. Ajeet Kumar Soni, Dr. Prashant Baredar,
and Dr. Satish Pal Singh Rajput

13.1 INTRODUCTION

In developing nations like India, post-harvest losses in food products are substantial. The main cause of such losses is improper storage and inadequate infrastructure. Food products are not adequately preserved to keep them fresh before they reach the marketplace. When the climate is hot and humid, wastage in production becomes more substantial. Agricultural goods preservation is important for their extended shelf life and for the sales and marketing of the perishable food products. It also increases the availability of food products after harvesting [1]. Preservation is nothing but to prevent the food items from loss or decay. Various preservation methods are used to prevent the loss of the quality or decay of agricultural products. In the case of perishable goods, to improve the shelf life, it is required to slow down or stop the bacterial decay [2]. In the present age, preservation of the agricultural goods such as vegetables and fruits has become more important as nonseasonal food products are commonly in demand.

The drying of food products involves the heat transfer along with the mass transfer from the food products. Heat is required to evaporate the surface moisture at the beginning phase of drying. Further to this energy is required to drive the moisture from the inner part of the product to its surface. So, a large amount of energy is required for the drying of agricultural goods. If this energy demand is fulfilled by using electrical energy as electricity, the combustion of a large amount of fossil fuels will be required to take place in conventional thermal power plants. It is known that it has been a major cause of environmental pollution since long decades. Also, the use of electricity for this purpose is not justifiable as around 23% of the population in India is still not getting electricity for their basic use. The other means of energy for drying, such as electrical generators using diesel as a fuel, produce greenhouse gases (GHG), which have become the cause of global warming. This also involves the transportation of diesel to village locations and causing further environmental issues.

DOI: 10.1201/9781003634737-13

Solar energy is abundantly available at most of the locations where the food products are produced. It can be easily and at a very low cost harnessed for the drying of food products [3]. So, solar drying is a promising alternative to all food drying methods. The literature shows that if the drying is done in a controlled manner, the food product could be kept with acceptable quality and standards, and the quality food will be sold at appreciable prices in the market. The Internet of Things (IoT) has been emerging as an advancement of information technology and by the use of advanced sensor technology, the solar dryer can be designed and developed with better efficiency [4]. Earlier, very few studies on solar drying were done by researchers in the late 1960s because of the use of fossil fuels in high-temperature drying through traditional methods. At that time, drying of agricultural goods by burning of fossil fuel was not very expensive [5]. Later, the industries involved in the drying realized the cost of drying by fossil fuels, and studies on using solar energy in drying started with the objective of solar dryer design.

Initially, some truss-shaped solar dryer design structures were made, which were able to be used for drying at low temperatures throughout the year. They were low-cost alternatives to the traditional high-temperature drying using fossil fuels [6]. In solar drying, specialized devices as solar dryers are used, which protect the agricultural products from damage due to insects, rain, and dust, as compared to direct sun drying. The solar dryers produce lesser spoilage of agricultural goods during the drying process and provide higher temperatures, lower moisture content, and relative humidity as compared to direct sun drying [7].

13.1.1 Need for Solar Dryers

The cause of the high proportion of wastage in agricultural production in India is chiefly due to the poor conditions of their storage and distribution. The lack of sufficient and suitable infrastructure is the reason in many cases. In India and similar countries, this wastage is about 10–40% of the production [8]. The other reason for the wastage of food products is due to the climatic conditions, improper handling, processing, and marketing of the produce. During the hot and humid environment, the quality is affected by bacterial decay in the goods.

To keep the food and fruit products fresh, various methods are conventionally adopted, such as refrigeration, drying, freezing, bottling, canning, and pickling. In the condition of nonavailability and high-cost involvement in the low-temperature maintenance and processing of food produce, solar drying has become an attractive and low-cost technique of preserving agricultural products. This provides an effective means of preserving the food items and thus reduces the post-harvest losses and the quality of the goods [9].

13.1.2 Solar Energy in Drying

The solar dryers are based on the solar energy received on the earth's surface. Some of the energy received from the sun at the top of the earth's atmosphere (1367 W/m^2), known as the solar constant dispersed from the earth's surface and scattered by the atmosphere [10]. The design consideration of solar dryers mainly includes the criterion of utilizing this remaining energy as available for drying to be effectively and efficiently utilized.

Figure 13.1 (a) shows the global horizontal irradiation as the sum of the direct and diffuse radiations for most of the countries. Figure 13.1 (b) shows

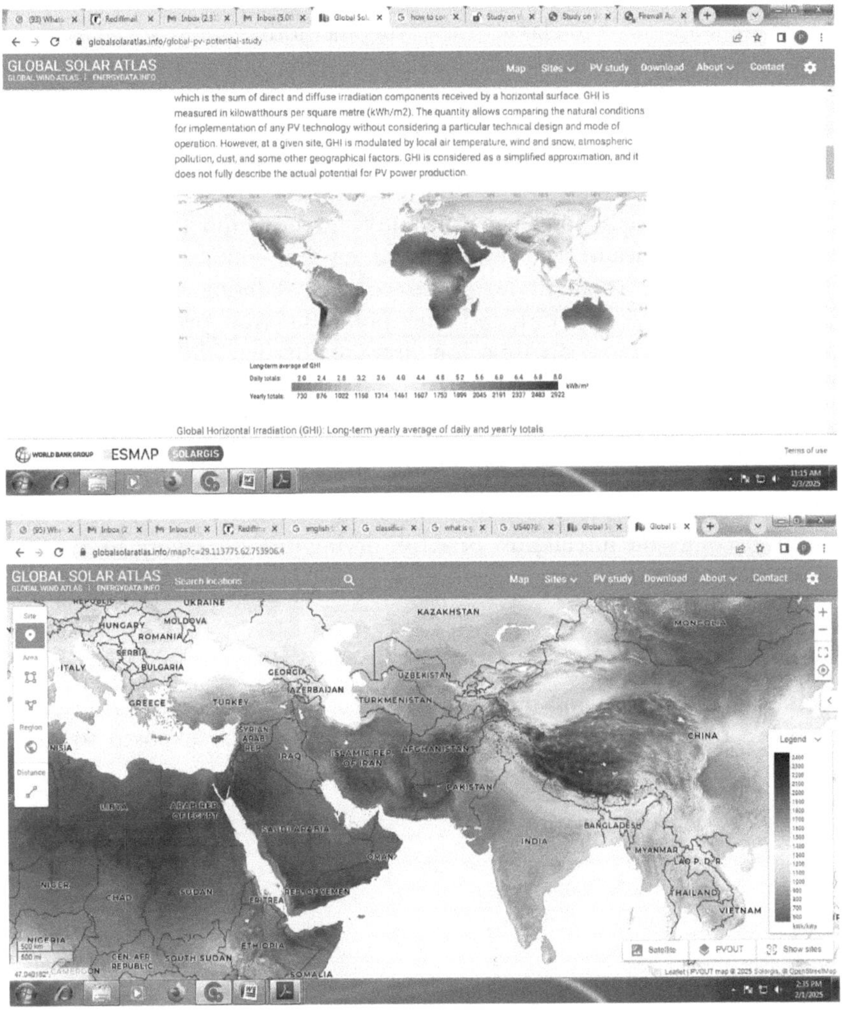

Figure 13.1 **(a–b)** Global horizontal irradiation.

the global horizontal irradiation focusing on India [11]. For the drying process, the heat required for 1 kg of water to be evaporated is 0.7 kWh, for the lower values of available solar radiation minimum of 700 W/m^2 energy from radiation is required for a solar dryer [12].

In India, the availability of solar radiation is good throughout the year. About 250–300 days in a typical year are sunny. The average solar radiation intensity in India is about 200 MW/km^2. The solar radiation intensity may be measured by methods including [13–15].

13.1.3 Drying Methods

There are many ways to classify the drying methods. Commonly, the drying methods may be broadly classified as conventional drying and solar drying. Conventional drawing uses fossil fuel or electricity for the generation of heat for drawing. Solar drying is commonly referred to as the drying of agricultural and food products by the use of solar energy [1]. The various methods of solar drying presently available have emerged over a long time with the advancement of technology. The principle of all the methods mainly involves the common modes of heat transfer, such as heat conduction, convection, and radiation. Direct sun drying offers economical drying, but the quality is deteriorated due to dust, dirt, and other pollutants from pets and rain [16].

Some of the methods are based on direct sun drying, while some use indirect solar energy for drying. A combination of these two is the mixed mode drying. A hybrid drying involves another source for heating, such as electricity, biomass, or LPG. If the air flow for the heat transfer is taking place naturally due to density difference, then it is called natural flow or passive type drying and if it is done forcefully using some mechanical means or blower then it becomes the forced flow or active drying method. Figure 13.2 shows the classification. The various methods have their own pros and cons.

13.1.4 Direct Solar Drying Method

In the direct solar drying method, the direct solar radiations are used as a source of heat energy. The agricultural products are exposed to these

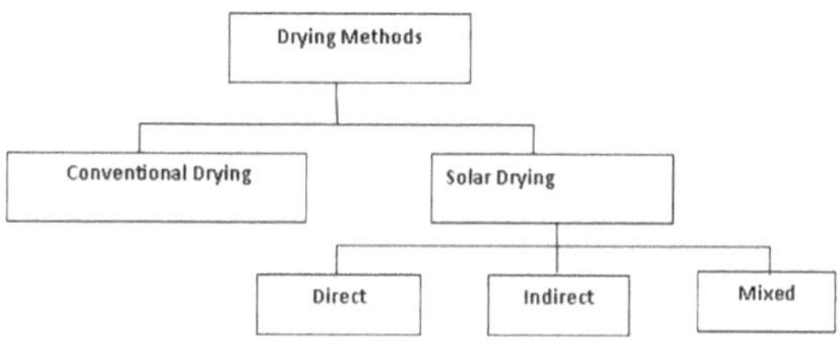

Figure 13.2 Classification of drying methods.

radiations. They are protected from the dust, dirt, and birds by separating the goods with a transparent cover of glass or plastic. The radiations falling over the transparent cover are transferred to the drying products, while some of them are reflected into the atmosphere. The radiations passing through the cover are also reflected to some extent, while the remainder is absorbed by the products to be dried.

13.1.5 Energy Storage in Solar Dryers

The energy received from the sun is stored in the solar dryers. Broadly, the method of energy storage can be classified as thermal energy storage and chemical energy storage. Thermal energy can be stored in liquids and solids as sensible heat by increasing their temperature. It is done when the available rich solar radiation can produce high temperatures [17]. The stored heat is released when the poor or no radiation is available and generates low temperatures. Water is most commonly used for heat storage due to its low cost and thermo-physical properties. Other materials used include rocks, bricks, and concrete materials.

In the latent heat storage method, the thermal energy can be stored in materials by a change in their phase. A special case of this type is solid–liquid phase change. It uses materials that have high latent heat storage with a lesser volume and almost constant temperature of melting and solidification. The various phase change materials can be classified as organic, inorganic paraffins, non-paraffins, and eutectic materials [18].

The thermal energy storage method is preferred over the chemical energy storage, where the heat pump or heat of chemical reaction is used [19]. Many researchers have worked with several heat storage materials and estimated the dryer's performance.

The study on the comparison of performances of a solar dryer using an air-heater-cum-rock-bed storage and a simple solar air heater used coriander food product. The test setup of the solar dryer consisted of a drying chamber, a rock-bed storage coupled with the drying chamber. It also used two similar double-pass, single-cover solar air heaters for the study. The results of this study show that about 27 sunshine hours were needed in three consecutive days to reduce the moisture content from 28.2% to 11.4% on a dry basis. Whereas, by using the solar dryer with rock-bed storage, only 18 sunshine and 13 off-sunshine hours in two consecutive days and nights were needed [20].

13.1.6 Advances in Design and Drying Methods

With the increasing research activities in solar drying technologies and dryer design, many methods and designs have been developed. There are several advantages and disadvantages of one dryer design and technology over the other. It has been the objective to increase the drying rate using the various phase-changing materials in hybrid designs. A lot of discussions can be made on various performance indicators of each design.

The advances in design and technology have taken place with the improvement in system parameters such as drying rate, moisture ratio, performance in terms of efficiency, energy consumption, and payback period. The experimental and CFD analysis have also contributed to such advances by providing valuable insights and scope of improvements in prior designs. It has been the objective of this study to focus on the broader aspect of such advancements and enlighten the recent technology that researchers are focusing upon with needed basic information [21, 22].

It is important to discuss the greenhouse drying method as a major advancement. Greenhouse drying is based on the greenhouse effect, which provides controlled heating inside an enclosed structure. The transparent cover allows the solar radiation to enter the enclosed space for drying. It was proposed by Huang as a greenhouse-bulk curing and drying system [23].

The greenhouse technology emerged to create a favorable environment primarily for use in horticulture [24]. Much research work on greenhouse dryers using solar energy has been underway and taking place for many years. As available on the Web of Science database, it was the first study on a prototype of a greenhouse dryer that was held in Yugoslavia. This dryer was used for drying mint and sage [25]. The designs are suitable for various load requirements. They have high initial cost, but it is justified for their ease in manufacturing, installation, and maintenance.

Technological advancements from continued research in greenhouse dryers have contributed many new designs and design guidelines. The designs for different applications and in both active and passive modes have been developed and in use in many countries of the world. Portable solar dryers have been developed for specific and mobile drying purposes [26]

The major advances in design and technology can be summarized as

1. Low-temperature drying of agricultural products and fruits and vegetables
2. Design improvements in dryers involving natural convection mode
3. In forced convection mode, energy provisions and hybrid designs
4. Mobile and portable designs to increase the application

A growing trend in the design and development of greenhouse solar dryers has been observed. Design modifications are based on the research outcomes of several researchers. Studies were conducted on research setups and by empirical studies. An empirical study on a greenhouse solar dryer to evaluate the drying efficiency of a greenhouse was conducted for winter and summer seasons. The results show similar trends in temperature and humidity range. For the winter, the temperature range obtained was 6.3°C–13.2°C, and humidity levels were in the range of 23%–40%. In summer, these ranges were 7.6°C–13.4°C for temperature and 26%–47% for the humidity levels [27–30].

A chief characteristic of a greenhouse effect is its ability to provide a suitable temperature range between 45°C and 60°C for drying of agricultural produce. Apart from this, it has a significant advantage of its effective use during whole of the year, providing economic feasibility in solar dryer applications [28]. Although the available research shows the influence of weather conditions on the drying time of various agricultural products, many fruits and vegetables of different shapes and sizes with different internal characteristics were dried using greenhouse solar dryer. It includes bananas, ginger, peppers, mangoes, garlic, etc. The results show the variations in drying times for different types of food [31–34].

13.1.7 Future Trends

With the recent technological advancements and involvement of network and remote technologies, the dryer design and technology of drying need to change. The mode of operation of solar dryers should be critically decided, and remote access of dryer systems by using modern internet-based technologies such as IoT will further enhance their importance in the food and agricultural production sector. This includes connected devices to work in the network [35–37]. Energy-efficient technologies and air distribution in the dryer may also be made more effective [17]. This technological involvement will have a substantial scope of modification in design and performance improvement in solar dryers.

13.1.8 Conclusion

Solar drying has been gaining significant popularity in recent years due to an increasing focus on the production of quality agricultural goods using solar energy and other environmental concerns. Conventionally, to retain the quality of the agricultural product, drying makes use of electricity, coal, fossil fuels, or direct solar rays. To reduce significant alteration in the quality of food products and to prevent the adverse effect of drying by conventional methods on the environment, advances have taken place in the design and drying technologies in recent years. As a promising alternative, solar energy is being used as a green energy in food drying. In the design and development of solar dryers, researchers have a great scope of work in recent times. Improving efficiency, maintaining strict and controlled thermal conditions, and continuous performance monitoring are the key issues to be focused upon.

An overview of solar dryer design and solar drying technology, with a specific focus on research outcomes based on experimental and numerical simulations, has been presented in the article. Various technological aspects, including design and development of solar dryers, operating modes, automation, energy requirements, drying performance, and hybrid drying, have been incorporated in this study. The outcome of this work shows that the solar dryers still have a substantial scope of modification in design and performance improvement. Their mode of operation should be critically decided, and remote access of these systems by using modern internet-based technologies will further enhance their importance in the food and agricultural production sector.

REFERENCES

[1] Abdullahi Y, Momoh M, Garba MM, Musa M (2013) Design and construction of an adjustable and collapsible natural convection solar food dryer. Int. J. Comput. Eng. Res. 3(6):1–8.

[2] Wankhade PK, Sapkal RS, Sapkal VS (2014) Design and performance evaluation of solar dryer. IOSR J. Mech. Civil Eng. 11:70–73.

[3] Wang W, Li M, Hassanien R, Hassanien E, Wang Y, Yang L (2018) Thermal performance of indirect forced convection solar dryer and kinetics analysis of mango. Appl. Therm. Eng. 134:310–321.

[4] Wadaskar A, Chouriya N, Alam T, Yadav S, Ninave V, Shaha R (2021) IoT based solar hybrid dryer. Int. J. Adv. Eng. Manag. 3(8):1680–1684.

[5] Kranzler GA, Bern CJ, Kline GL, Anderson ME (1980) Grain drying with supplemental solar heat. Am. Soc. Agric. Eng. 214–217. http://dx.doi.org/10.13031/2013.34557.

[6] Phillips AL (1965) Drying coffee with solar-heated air. Sol. Energy 9(4):213–216. http://dx.doi.org/10.1016/0038–092x(65)90051–4.

[7] Umogbai VI, Iorter HA (2013) Design, construction and performance evaluation of a passive solar dryer for maize cobs. Afr. J. Food Sci. Technol. 4(5):110–115.

[8] Rosa Rolle S (2006) *Postharvest management of fruit and vegetables in the Asia-Pacific region*. Asian Productivity Organization, Tokyo, Japan.

[9] Heydari A, Forati M, Khatam SM (2020) Thermal performance investigation of a hybrid solar air heater applied in a solar dryer using thermodynamic modeling. J. Therm. Eng. 79:715–730.

[10] Garg HP, Prakash J (1997) *Solar energy fundamentals and applications*. Tata McGraw Hill, New Delhi, India.

[11] Global Solar Atlas (2025) *Global horizontal irradiation obtained from the "Global Solar Atlas 2.0", a free, web-based application* [Accessed on 25 January 2025]. https://globalsolaratlas.info/map.

[12] de Dios CA (1996) *Secado de granos y secadoras.* 1st ed. Hemisferio Sur Editorial.
[13] Hemaiswarya S, Raja R, Carvalho IS, Ravikumar R, Zambare V, Barh D (2012) An Indian scenario on renewable and sustainable energy sources with emphasis on algae. Appl. Microbiol. Biotechnol. 96:1125–1135. https://doi. org/10.1007/s00253-012-4487-0.
[14] Reddy VS, Kaushik SC, Panwar NL (2013) Review on power generation scenario of India. Renew. Sustain. Energy Rev. 18:43–48. https://doi.org/10.1016/j.rser.2012.10.005.
[15] Hemaiswarya S, Raja R, Carvalho IS, Ravikumar R, Zambare V, Barh D (2012) An Indian scenario on renewable and sustainable energy sources with emphasis on algae. Appl. Microbiol. Biotechnol. 96:1125–1135. https://doi. org/10.1007/s00253-012-4487-0.
[16] Hii CL, Jangam SV, Ong SP, Mujumdar AS (eds) (2012) *Solar drying: Fundamentals, applications and innovations.* TPR Group Publication, Singapore.
[17] Bal LM, Satya S, Naik SN, Meda V (2011) Review of solar dryers with latent heat storage systems of agricultural products. Renew. Sustain. Energy Rev. 15:876–880.
[18] Sharma A, Tyagi VV, Chen CR, Buddhi D (2009) Review on thermal energy storage with phase change materials and applications. Renew. Sustain. Energy Rev. 13:318–345.
[19] Bal LM, Satya S, Naik SN (2010) Solar dryer with thermal energy storage systems for drying agricultural food products: A review. Renew. Sustain. Energy Rev. 14:2298–2314.
[20] Chauhan PM, Choudhury C, Garg HP (1996) Comparative performance of coriander dryer coupled to solar air heater and solar air-heater-cum-rockbed storage. Appl. Therm. Eng. 16:475–486.
[21] Janjaia S, Srisittipokakuna N, Balab BK (2008) Experimental and modelling performances of a roof-integrated solar drying system for drying herbs and spices. J. Energy 33:91–103.
[22] Ohunakin OS, Adaramola MS, Oyewola OM, Fagbenle RO (2013) Correlations for estimating solar radiation using sunshine hours and temperature measurement in Osogbo, Osun State. Front. Energy 7(2):214–222.
[23] García J, Too YC, Padilla RV, Beath A, Kim JS, Sanjuan ME (2018) Dynamic performance of an aiming control methodology for solar central receivers due to cloud disturbances. Renew. Energy 121:355–367. https://doi.org/10.1016/j.renene.2018.01.019.
[24] Bal LM, Satya S, Naik S (2010) Solar dryer with thermal energy storage systems for drying agricultural food products: A review. Renew. Sustain. Energy Rev. 14(8):2298–2314. http://dx.doi.org/10.1016/j.rser.2010.04.014.
[25] Műller J, Reisinger G, Kisgeci J, Kotta E, Tesic M, Mühlbauer W (1989) Development of a greenhouse-type solar dryer for medicinal plants and herbs. Sol. Wind Technol. 6(5):523–530. http://dx.doi.org/10.1016/0741–983X(89)90086–6.
[26] Mhd Safri N, Zinuddin Z, Modh A, Fudholi A, Zulkifle I, Hafidz R (2020) Temperature performance of a portable solar greenhouse dryer with various collector design. Sains. Malays. 49(10):2539–2545. http://dx.doi.org/10.17576/jsm-2020-4910-19.

[27] Mazzeo D, Baglivo C, Panico S, Congedo PM (2021) Solar greenhouses: Climates, glass selection, and plant well-being. Sol. Energy 230:222–241. http://dx.doi.org/10.1016/j.solener.2021.10.031.
[28] Bal LM, Satya S, Naik S (2010) Solar dryer with thermal energy storage systems for drying agricultural food products: A review. Renew. Sustain. Energy Rev. 14(8):2298–2314. http://dx.doi.org/10.1016/j.rser.2010.04.014.
[29] El-Sebaii A, Shalaby S (2012) Solar drying of agricultural products: A review. Renew. Sustain. Energy Rev. 16(1):37–43. http://dx.doi.org/10.1016/j.rser.2011.07.134.
[30] Tiwari S, Swaminathan M, Eashwar SS, Harender, Singh DB (2022) Performance enhancement of the photovoltaic system with different cooling methods. Environ. Sci. Pollut. Res. 29(30):45107–45130. https://link.springer.com/article/10.1007/s11356-022-20330-x.
[31] Kanouni B, Badoud AE, Mekhilef S, Bajaj M, Zaitsev I (2024) Advanced efficient energy management strategy based on state machine control for multi-sources PV-PEMFC-batteries system. Sci. Rep. 14(1):7996. https://doi.org/10.1038/s41598-024-58785-2.
[32] Tajjour S, Chandel SS, Alotaibi MA, Malik H, Márquez FP, Afthanorhan A (2023) Short-term solar irradiance forecasting using deep learning techniques: A comprehensive case study. IEEE Access 11:119851–119861. http://dx.doi.org/10.1109/ACCESS.2023.3325292.
[33] Molu RJ, Tripathi B, Mbasso WF, Naoussi SR, Bajaj M, Wira P, Blazek V, Prokop L, Misak S (2024) Advancing short-term solar irradiance forecasting accuracy through a hybrid deep learning approach with Bayesian optimization. Results Eng. 23:102461. https://doi.org/10.1016/j.rineng.2024.102461.
[34] Singh AR, Kumar RS, Bajaj M, Khadse CB, Zaitsev I (2024) Machine learning-based energy management and power forecasting in grid-connected microgrids with multiple distributed energy sources. Sci. Rep. 14(1):19207. https://doi.org/10.1038/s41598-024-70336-3.
[35] Yadav D, Awasthi MK, Kumar A (2025) *Next generation renewable thermal energy harvesting, conversion & storage technologies*. Elsevier.
[36] Kumar A, Singh VP, Meena CS, Dutt N (2023) *Thermal Energy Systems: Design, Computational Techniques and Applications*. CRC Press, Boca Raton, FL. https://doi.org/10.1201/9781003395768.
[37] Tomar P, Chauhan P, Kumar A, Shukla AK, Macedo VD, Mishra BP, Lakkimsetty NR, Manjunath TC (2024) Numerical investigation of thermal performance enhancement of solar reservoir using flash cycle. J. Adv. Res. Fluid Mech. Therm. Sci. 123(1):197–221. https://doi.org/10.37934/arfmts.123.1.197221.

Chapter 14

Wind Solar Hydrogen Hybrid Energy Generation System Prospects

Saksham Gaur, Shresth Saini, and Meeta Sharma

14.1 INTRODUCTION

14.1.1 Background on Renewable Energy Systems

Renewable energy has emerged as a cornerstone in addressing critical challenges such as climate change, energy security, and sustainable development. Over the past few decades, the way energy is produced has undergone a significant transformation, with an increasing emphasis on renewable resources like hydropower, geothermal, wind, and solar energy. Unlike fossil fuels, which are finite and contribute substantially to greenhouse gas emissions, renewable energy sources offer an inexhaustible supply with minimal environmental impact. Recognizing these advantages, governments, businesses, and individuals worldwide are prioritizing renewable energy to reduce reliance on fossil fuels and mitigate the risks of climate change (Abu-Rayash and Dincer, 2021, 2023). Today, the energy sector faces three major challenges: the urgent need for cleaner energy, a stable and consistent supply, and the capacity to meet the growing global energy demand. Historically, energy systems have depended on fossil fuels such as coal, oil, and natural gas. While these sources provide reliable power and high energy density, they come at a steep environmental cost. Fossil fuel combustion has been a significant contributor to global warming, with carbon dioxide (CO_2) being the primary greenhouse gas. The transition to renewable energy is widely regarded as a vital solution to address these environmental concerns (Badwawi et al., 2015; Bekele and Palm, 2010).

Among renewable technologies, solar and wind power systems have achieved prominence due to their efficiency, scalability, and cost-effectiveness. Advances in wind turbines and solar photovoltaic (PV) panels have made them competitive alternatives to conventional energy sources (Chávez-Ramírez et al., 2013; Couto and Estanqueiro, 2023). Solar energy harnesses sunlight, while wind energy captures the kinetic energy of moving air, both of which are abundant and accessible across diverse geographies. However, these sources are inherently variable; solar energy depends on weather and

DOI: 10.1201/9781003634737-14

daylight conditions, while wind energy is influenced by seasonal and regional wind patterns (Ding et al., 2019; Goyal et al., 2017).

Globally, the adoption of solar PV systems and wind turbines has accelerated significantly. By 2023, solar and wind energy together accounted for over 45% of the world's renewable energy generation. Europe has led advancements in offshore wind energy, while countries like China, the United States, and India have spearheaded large-scale solar and wind farm installations. Despite their immense potential, the unpredictability of these resources presents a significant challenge, particularly when they constitute a large share of the energy mix. Variations in sunlight and wind conditions can cause imbalances between energy supply and demand, complicating grid stability (Gu et al., 2020; Gupta et al., 2007; Hirose and Matsuo, 2011).

To address these challenges, energy storage technologies such as batteries and hydrogen fuel cells have become indispensable. These systems store excess energy produced during peak production periods and release it when output drops, thus stabilizing the grid. Lithium-ion batteries are frequently paired with solar and wind power to store surplus energy efficiently. Similarly, hydrogen fuel cells, which store energy as hydrogen gas and convert it back into electricity when needed, are gaining popularity for their potential in long-term storage (Jenkins et al., 2019; Kabalci, 2013; Khare et al., 2016; Liu et al., 2020).

In addition to storage, hybrid renewable energy systems have emerged as an innovative solution to the variability of renewable generation. These systems combine two or more renewable sources, such as solar and wind, with energy storage components, offering a more reliable and balanced energy supply (Liu et al., 2024; Nasser et al., 2022; Rehman et al., 2018). For instance, while solar energy peaks during the day, wind energy often excels at night or during cooler seasons. By working in tandem, hybrid systems can complement each other's strengths, ensuring continuous energy production and enhancing overall system efficiency (Temiz and Dincer, 2022; Temiz and Javani, 2020).

14.2 IMPORTANCE OF HYBRID WIND-SOLAR SYSTEMS

Hybrid renewable energy systems, particularly those combining solar and wind power, have gained traction due to their ability to overcome the limitations of single-source energy generation (Yang et al., 2007, 2009; Singh et al., 2024). By leveraging the complementary nature of solar and wind energy, hybrid systems deliver a more consistent and predictable energy output. Solar power typically peaks during sunny days, while wind energy generation is higher at night or during windy seasons. This complementary behavior ensures energy availability across various conditions, reducing the need for backup power from fossil fuels (Jamil et al., 2023; Chakrabarty et al., 2017).

Hybrid systems are particularly effective in areas with high solar irradiation and consistent wind speeds, where they can provide a stable and reliable energy supply. For example, wind energy can offset the reduced solar output at night or during cloudy weather, while solar energy can compensate for low wind speeds during calm conditions. This synergy minimizes fluctuations in energy production and enhances the reliability of the overall system. Energy storage further strengthens the reliability of hybrid systems. Batteries and hydrogen fuel cells are critical for storing excess energy generated during peak periods for use during lower production times. For instance, a hybrid system equipped with battery storage can store surplus solar energy produced during the day for nighttime use. Similarly, wind energy generated during high wind speeds can be stored for periods of calm. This ability to store and release energy on demand ensures that hybrid systems can deliver steady power even when renewable generation is temporarily unavailable (Bhandari et al., 2023; Rekioua et al., 2023; Kanouni et al., 2024).

One of the key advantages of hybrid wind-solar systems is their ability to enhance grid stability and reduce reliance on fossil fuels as backup energy sources. Traditional power grids often rely on fossil fuel-based plants to provide "baseload" electricity, a steady supply to meet consistent demand. Hybrid systems, with their integrated storage capabilities, can reduce this dependence by supplying electricity during periods of low renewable generation. This not only lowers operational costs associated with maintaining fossil fuel plants but also significantly reduces carbon emissions (Almutairi et al., 2023; Tiwari et al., 2022).

Hybrid systems also offer substantial cost benefits by optimizing resource utilization. By integrating solar and wind energy within a single system, existing land and infrastructure are utilized more efficiently. For example, installing both wind turbines and solar panels in the same location maximizes energy production per unit area. Shared infrastructure, such as inverters and grid connections, further reduces installation and maintenance costs, making hybrid systems more economically viable than standalone renewable systems (Shankar, 2023; Badri et al., 2023). The adaptability of hybrid systems makes them suitable for a wide range of applications, from small-scale local installations to large utility projects. In remote or off-grid areas, hybrid systems provide a reliable energy source without the need for extensive grid infrastructure. They are increasingly used in rural regions, islands, and developing countries to power homes, businesses, and communities, reducing dependence on costly and environmentally harmful diesel generators (Verma et al., 2024; Bansal et al., 2024).

Beyond cost and reliability, hybrid systems enhance energy resilience. As energy infrastructure becomes more vulnerable to disruptions from natural disasters, extreme weather, and cyber-attacks, the need for robust systems has grown. Hybrid systems, with their decentralized generation and storage capabilities, offer a more resilient solution. During grid outages, they can provide power to critical services like hospitals and emergency facilities, ensuring continuity during crises (Kumar et al., 2024).

Finally, hybrid wind-solar systems play a crucial role in meeting the growing global energy demand. The International Energy Agency (IEA) predicts a 30% increase in energy demand by 2040, driven by population growth, urbanization, and economic development. To meet this demand while reducing greenhouse gas emissions, large-scale deployment of renewable energy is essential. Hybrid systems, with their ability to deliver reliable and uninterrupted power, are indispensable in achieving this goal and transitioning to a sustainable energy future (Meena et al., 2024; Kumar et al., 2023; Yadav et al., 2025).

14.3 LITERATURE REVIEW

The transition of the world's energy sources to renewables is essential for combating climate change and advancing sustainable development. With their own benefits, solar and wind energy have become the major renewable energy sources. By integrating them into hybrid systems, they overcome the drawbacks of using separate energy sources and improve energy efficiency and reliability.

14.4 KEY OBSERVATIONS

14.4.1 Solar Energy Systems

Photovoltaic (PV) panels and concentrated solar power (CSP) systems are the primary technologies for harvesting solar energy. PV panels utilize semiconductors to directly convert sunlight into electricity, offering high efficiency, modular scalability, and minimal maintenance post-installation (Amoussou, 2024). However, their performance can decline under cloudy conditions or at night, affecting consistent energy generation. In contrast, CSP systems use mirrors or lenses to concentrate sunlight, generating heat that is converted into electricity via steam turbines. This technology excels in sunny regions, where thermal energy storage enables power generation even after sunset (Molu, 2024). Additional solar solutions, such as floating solar farms and solar thermal collectors, provide tailored options for specific applications.

14.5 WIND ENERGY SYSTEMS

Wind energy, primarily captured using horizontal-axis wind turbines (HAWT) and vertical-axis wind turbines (VAWT), plays a critical role in renewable energy production. HAWTs, known for their efficiency, are commonly employed in both onshore and offshore installations (Sahri et al., 2023). While typically less efficient, VAWTs can capture wind from any direction, making them suitable for urban environments. Together, these technologies significantly enhance energy production, particularly in areas with favorable wind conditions.

14.6 ENERGY STORAGE TECHNOLOGIES IN HYBRID SYSTEMS

Energy storage is crucial in bridging the gap between intermittent energy output and continuous energy consumption in hybrid systems. Various technologies have been explored to address this challenge. Punyavathi et al. (2024) emphasized the importance of storage in maintaining standalone hybrid systems' reliability, particularly when integrated with load-following algorithms and maximum power point tracking (MPPT) techniques. Kabalci (2013) introduced ultra-capacitors as a novel stabilization method, preventing battery overcharging and ensuring seamless energy transitions.

Das et al. (2022) evaluated vehicle to grid and grid to vehicle strategies, highlighting their potential to improve power quality and voltage stability. Their research also underscored the importance of inverter controllers and compensators in addressing voltage and frequency fluctuations in grid-connected systems. Liu et al. (2024) examined the cost-effectiveness of thermal storage systems and batteries, emphasizing the need to align storage technologies with the complementary nature of solar and wind energy to optimize efficiency.

A notable example is the wind-photovoltaic-hydrogen system developed by Temiz and Dincer (2022). This system provided energy autonomy for a community of 5,300 residents while enabling sustainable hydrogen production for long-term storage. With a maximum hydrogen production capacity of 653.9 tons per year and a consumption rate of 636.3 tons, the system demonstrated exceptional efficiency, particularly during peak cooling months.

14.7 CONTROL STRATEGIES AND OPTIMIZATION TECHNIQUES

Effective control and optimization strategies are essential for the stability and efficiency of hybrid renewable energy systems. Bhatt et al. (2024) emphasized the role of scalable control systems incorporating battery storage and diesel generators in rural electrification projects. Their findings demonstrated how improved control systems enhance energy supply reliability, particularly in remote areas.

Khare et al. (2016) explored Particle Swarm Optimization (PSO) techniques for managing distributed generation in hybrid systems. These algorithms effectively minimized power supply losses and increased overall system efficiency. Liu et al. (2024) advanced this concept with load-tracking algorithms that dynamically adjust energy distribution and storage based on real-time demand. These techniques significantly reduced operating costs and improved system performance. Research by Temiz and Dincer (2022) demonstrated an 83% reduction in the levelized cost of electricity (LCOE) for hybrid systems compared to solar-only setups. The hybrid system achieved a maximum overall efficiency of 42.63%, underscoring the transformative potential of advanced optimization techniques.

14.8 SOCIO-ECONOMIC AND ENVIRONMENTAL IMPACTS

Hybrid renewable energy systems offer substantial socio-economic and environmental benefits, particularly for communities reliant on fossil fuels. Bekele and Palm (2010) highlighted the potential of hybrid systems to improve energy access in rural Ethiopia, noting significant cost reductions and environmental benefits from increasing renewable energy's share to 51%. This transition reduced diesel dependency and greenhouse gas emissions.

Rekioua et al. (2023) emphasized the economic advantages of incorporating energy storage into hybrid wind-solar systems for rural electrification. Their findings revealed that hybrid systems could deliver reliable power at lower costs than traditional setups, fostering rural development. The socio-economic impact of Temiz and Dincer's (2022) hybrid wind-PV-hydrogen system was particularly notable. Serving a 5,300-person community, the system achieved 100% energy autonomy and drastically reduced energy costs. With a competitive LCOE of CAD$0.128/kWh and a hydrogen production cost of CAD$4.6/kg, the system exemplified the economic viability of hydrogen storage in hybrid setups. Additionally, its ability to meet all energy needs—including heating, cooling, and electricity—makes it a compelling model for sustainable community development.

Rekioua et al. (2023) demonstrated the environmental benefits of hybrid systems through a simulation of a concentrating solar power (CSP) and wind turbine system. Their findings highlighted the potential for hybrid systems to achieve energy autonomy, reduce greenhouse gas emissions, and provide sustainable energy solutions for remote or island communities.

14.9 KEY FINDINGS

14.9.1 Hybrid Configurations

The literature highlighted the potential of solar-wind hybrid systems in addressing the intermittency issues inherent in renewable energy sources. Solar and wind energy sources complement each other, with solar output typically peaking during the day and wind power often stronger during the evening or night. This complementarity enables a more stable energy generation profile compared to standalone systems. However, despite these advantages, several limitations were observed. One significant gap was the underutilization of hydrogen-based storage systems. Hydrogen storage offers the advantage of storing energy over longer durations, making it ideal for seasonal or extended use cases. Yet, many studies have underutilized this potential, opting for conventional battery storage or standalone hydrogen systems. The limited adoption of hydrogen-based storage leaves a significant gap in the development of more efficient hybrid configurations (Figure 14.1).

Additionally, there was an unbalanced integration of storage solutions, with most studies focusing on either short-term battery storage or standalone hydrogen systems. This singular focus often failed to exploit the complementary strengths of batteries, which offer high efficiency and fast response, and hydrogen, which provides long-term storage and higher energy density.

14.10 ENERGY STORAGE TECHNOLOGIES

Energy storage systems are crucial for managing the natural variability of solar and wind power. While both batteries and hydrogen systems have been extensively discussed, specific challenges and gaps in their application persist. Lithium-ion batteries, known for their superior charge-discharge efficiency, extended lifecycle, and suitability for short-term energy storage, were widely acknowledged. However, challenges such as limited capacity for long-term storage and susceptibility to performance degradation under deep discharge conditions remain.

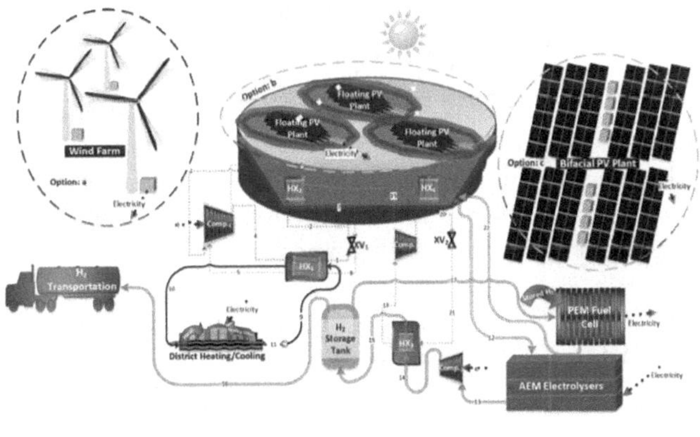

Figure 14.1 Layout of the proposed system by Temiz, M. and Dincer, I. (Temiz and Dincer, 2022).

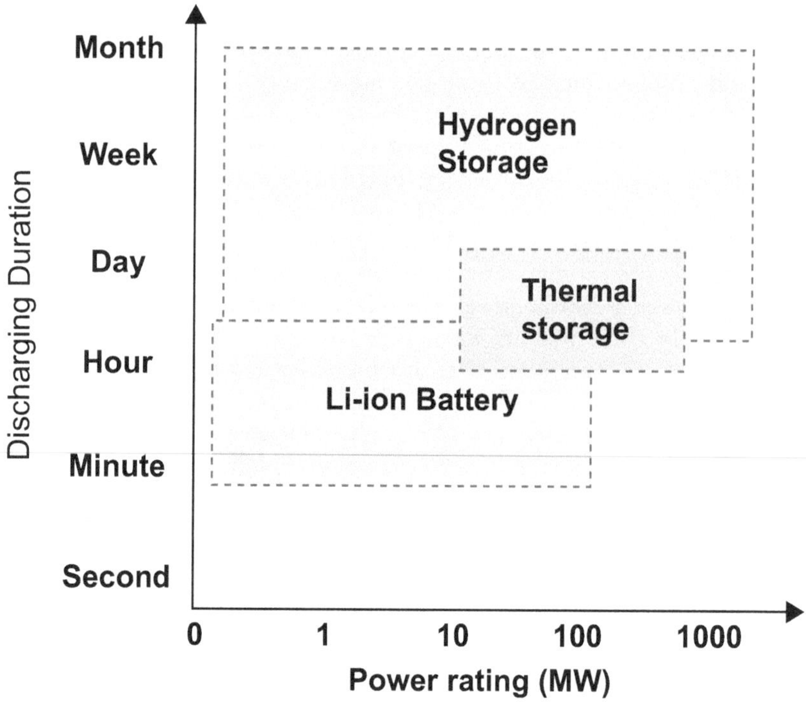

Figure 14.2 Power rating and discharging duration (Liu et al., 2024).

On the other hand, hydrogen storage systems show promise for long-term energy storage applications by leveraging excess renewable energy to produce hydrogen through electrolysis. Hydrogen can be stored for extended periods and converted back into electricity using fuel cells when needed. Despite these advantages, few studies have provided clear strategies for integrating hydrogen systems into hybrid configurations, leaving a gap in the practical implementation of these systems.

Furthermore, advanced optimization techniques like Particle Swarm Optimization (PSO) were explored to improve energy distribution and minimize losses. While these techniques have potential, they were not widely applied to systems with diverse energy sources, leaving a gap in the efficient utilization of resources in hybrid setups (Figure 14.2).

14.11 CONTROL STRATEGIES

Efficient control systems are essential for managing the complexities of hybrid energy systems. The literature revealed limitations in existing control approaches, particularly in mode-specific systems. Many studies focused solely on either grid-connected or standalone systems, neglecting the challenges of transitioning seamlessly between these modes. This is particularly critical for ensuring uninterrupted power supply during grid failures or fluctuations, a need that is often overlooked.

Additionally, the literature highlighted the necessity for sophisticated energy management systems (EMS) that integrate real-time monitoring, predictive analytics, and adaptive algorithms. These systems must dynamically balance energy generation, storage, and distribution to meet fluctuating demand. However, many existing solutions lacked the advanced functionality required for real-time decision-making and adaptive responses to changing conditions, underscoring a significant gap in current control strategies.

14.12 ENVIRONMENTAL AND SOCIO-ECONOMIC IMPACTS

The environmental benefits of hybrid systems, particularly in reducing greenhouse gas emissions and reliance on fossil fuels, were widely acknowledged in the literature. Hybrid systems also demonstrated the potential to provide energy autonomy, especially in remote or off-grid locations. However, there were several noted limitations, particularly in terms of scalability and cost-effectiveness.

While small-scale implementations of hybrid systems have proven effective, their scalability for larger, community-level, or industrial applications has not been adequately explored. Scaling up these systems introduces challenges related to cost, resource availability, and system efficiency, all of which remain largely unaddressed in many studies. Furthermore, the economic viability of hybrid systems, particularly those incorporating hydrogen storage, was underexplored. These systems often face high upfront costs and operational expenses, and there is limited research on strategies to optimize costs while maintaining performance. This gap in the literature highlights the need for more comprehensive studies on the economic and practical feasibility of large-scale hybrid systems (Figure 14.3).

14.13 GAPS AND OPPORTUNITIES

The literature review identified critical gaps in the design, integration, and optimization of hybrid renewable energy systems. Specifically, the following limitations were observed:

- **Lack of Balanced and Integrated Energy Storage Solutions:** Many studies focused on either batteries or hydrogen storage systems but rarely combined in both a balanced and an integrated manner. This

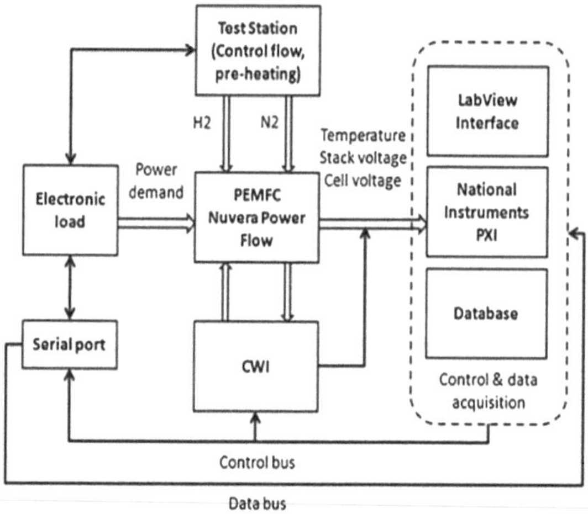

Figure 14.3 Schematic representation of PEMFC (Chávez-Ramírez et al., 2013).

unbalanced approach often failed to exploit the complementary strengths of these storage technologies, leaving a significant gap in hybrid system development.

- **Limited Application of Advanced Optimization Techniques:** Despite the potential of advanced optimization techniques such as Particle Swarm Optimization (PSO), these methods were not widely applied to hybrid systems with diverse energy sources. The limited use of such techniques for efficient energy distribution and minimization of losses presents an opportunity for further research and application.
- **Insufficient Focus on Seamless Control Strategies:** A key challenge identified was the lack of seamless control strategies for transitioning between grid-connected and standalone modes. Most existing studies failed to address the need for efficient control systems capable of dynamically managing this transition, which is essential for ensuring a continuous power supply during grid disruptions.
- **Inadequate Exploration of Scalability and Cost-Effectiveness:** The scalability of hybrid renewable energy systems for large-scale applications and their economic viability were not sufficiently explored in many studies. Most research focused on small-scale systems, neglecting the challenges and opportunities associated with scaling up hybrid configurations for industrial or community-level energy demands.

Addressing these gaps presents a significant opportunity to develop more reliable, efficient, and scalable hybrid renewable energy systems capable of meeting the growing global demand for clean, sustainable energy. The

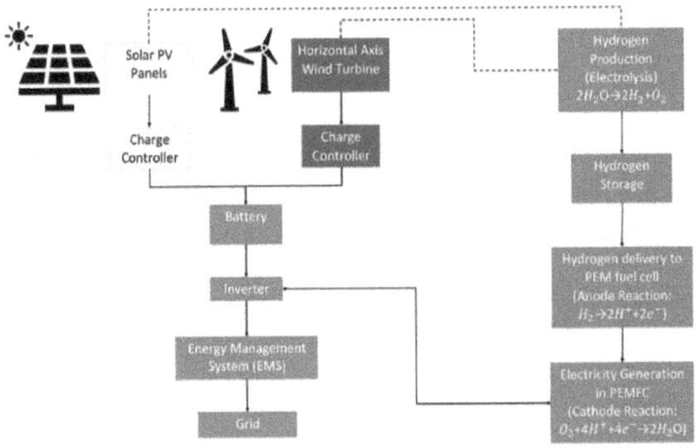

Figure 14.4 Proposed layout of the hybrid system.

findings from this literature review provide the foundation for the innovations and contributions presented in this work (Figure 14.4).

14.14 FUTURE WORK

Future research can build upon the findings of this study by further addressing gaps and limitations in hybrid renewable energy systems. Expanding on the innovative approach to design, integration, and optimization, future work can explore advanced configurations of hybrid energy systems, incorporating emerging technologies such as next-generation solar photovoltaic (PV) modules, enhanced wind turbine designs, improved Proton Exchange Membrane (PEM) fuel cells, and next-gen battery storage solutions. Additionally, research can focus on optimizing system performance through artificial intelligence-driven predictive analytics and real-time energy management strategies to enhance reliability, efficiency, and scalability.

14.15 INSIGHTS FROM THE LITERATURE REVIEW

Future research can address several key challenges identified in this study regarding hybrid renewable energy systems. One critical area for exploration is the configuration of hybrid systems, particularly solar-wind hybrids, with a stronger emphasis on achieving a balanced integration of energy storage solutions. While hydrogen-based storage has shown promise, further investigation is needed to optimize its combination with batteries for both long-term and short-term energy requirements, ensuring enhanced reliability and efficiency.

Advancements in energy storage technologies also present a valuable avenue for future work. While batteries are widely recognized for their high charge-discharge efficiency and rapid response times, their limitations in long-term storage necessitate further innovation. Future research can focus on developing robust implementation strategies for hydrogen storage within hybrid systems, addressing existing gaps and improving overall energy resilience.

Another important aspect is the enhancement of control strategies. Existing approaches primarily cater to either grid-connected or standalone modes, often neglecting the need for seamless transitions. Future work should focus on developing advanced energy management systems (EMS) with real-time monitoring and adaptive control capabilities to ensure efficient and dynamic system operation under varying conditions.

Lastly, addressing scalability and cost-effectiveness remains a crucial challenge. Most studies to date have focused on small-scale systems, limiting their applicability to large-scale hybrid configurations. Future research should prioritize cost-efficient solutions and scalable designs that can meet industrial and community-level energy demands, paving the way for widespread adoption of hybrid renewable energy systems.

14.16 ADVANCES

Future research can build upon the advancements introduced in this study to further refine hybrid renewable energy systems. One key area for exploration is the enhancement of dual energy storage integration. While this study combined lithium-ion batteries for short-term fluctuations with PEM fuel cells for long-term energy security, future work can optimize this synergy by incorporating emerging battery chemistries and advanced hydrogen storage techniques to further improve efficiency and reliability. Additionally, research can focus on optimizing hydrogen production strategies from surplus solar PV and wind energy to ensure even greater resource utilization.

Expanding on system design, future studies can investigate more advanced energy management systems (EMS) capable of adaptive, AI-driven decision-making for seamless transitions between grid-connected and islanding modes. Enhancements in PEM fuel cell systems, particularly in water management and membrane hydration mechanisms, can be explored to boost efficiency and scalability under varying operational conditions.

Optimization techniques can also be further developed to enhance energy distribution and minimize losses. While this study utilized Particle Swarm Optimization (PSO), future research can explore hybrid optimization algorithms that combine machine learning approaches with traditional methods to improve energy flow, reduce the levelized cost of electricity (LCOE), and increase overall system performance. Additionally, integrating more sophisticated dynamic load-tracking algorithms can refine real-time energy dispatch, ensuring optimal utilization of solar and wind energy resources.

Future work can also advance control strategies through next-generation EMS frameworks that leverage real-time monitoring, predictive analytics, and decentralized control mechanisms to optimize energy generation, storage, and distribution. The integration of adaptive algorithms can improve system interaction with the grid, enhancing resilience and overall reliability. Finally, scalability and economic feasibility remain critical areas for further research. Future studies should focus on refining modular system designs to enable flexible adaptation for applications ranging from residential microgrids to utility-scale installations. Further economic analysis is needed to explore innovative financing models and policy incentives that can lower costs, particularly in reducing LCOE and hydrogen production expenses, making hybrid renewable energy systems more accessible for widespread deployment.

14.17 UNIQUE CONTRIBUTIONS

Future research can build upon the key contributions of this study to further advance hybrid renewable energy systems. One important area for exploration is the refinement of balanced energy storage integration. While this study demonstrated the effectiveness of combining lithium-ion batteries and PEM fuel cells for short-term and long-term storage, future work can investigate novel energy storage technologies, such as solid-state batteries and advanced hydrogen storage methods, to enhance efficiency and energy density.

Further research can also focus on improving the seamless operational flexibility of hybrid systems. While this study introduced dual-mode operation (grid-connected and islanding), future work can develop intelligent transition mechanisms that leverage AI-driven predictive analytics for even smoother mode shifts. Enhancing system resilience in areas with unreliable grid infrastructure or frequent disruptions remains a critical area for exploration.

Scalability and adaptability of hybrid renewable energy systems also require further study. Future research should explore the development of modular system architectures optimized for different energy demand profiles, ensuring cost-effective deployment for both residential microgrids and large-scale industrial applications. Investigating the impact of emerging policy frameworks and financial models on system affordability and adoption could further support widespread implementation. Sustainability and cost-effectiveness will continue to be key drivers of hybrid energy system innovation. Future work can examine more sophisticated optimization techniques that enhance the synergy between renewable energy sources while minimizing environmental impact. Additionally, economic modeling can explore pathways to further reduce the levelized cost of electricity (LCOE) and hydrogen production expenses, making hybrid energy solutions more viable for global deployment.

By addressing these areas, future research can build upon the foundation established in this study, driving the next generation of hybrid renewable energy systems that are more reliable, efficient, and scalable.

14.18 CONCLUSION

The study highlights the possibility of a hybrid renewable energy system that combines solar photovoltaics, wind turbines, hydrogen fuel cells, and battery storage to meet global energy concerns. The literature study emphasizes how vital it is to move away from fossil fuels due to the world's growing energy needs and the need to slow down climate change. Despite their significant growth in the global energy scene, intermittent sources like solar and wind energy are not able to supply power continuously.

The examined studies highlight the significance of integrating storage options with renewable energy sources, especially hydrogen fuel cells, which are still underutilized in present applications. The long-term storage capacity that hydrogen provides as an energy transporter balances out the daily variations in solar and wind energy generation. Therefore, the integration of these technologies is vital for building a dependable, robust, and sustainable energy system. Building on this, the present work suggests a 50 MW hybrid system that maximizes energy generation and storage to stabilize the grid and provide a constant power supply. This hybrid system can drastically lower dependency on fossil fuels by combining the high efficiency of solar PV systems, the dependability of wind turbines, and the storage capacity of hydrogen fuel cells. To further improve system stability, battery storage is also essential for managing short-term energy demands. The study concentrates on sophisticated arrangements of every element, underscoring the vital function of hydrogen in making up for inadequate renewable energy production when there is not enough wind or sunlight.

The study's mathematical modeling provides in-depth information about the hybrid system's energy output, peak power, system efficiency, and reduction of CO_2 emissions. The system's ability to generate significant amounts of energy while offsetting greenhouse gas emissions is demonstrated by the equations presented for the energy outputs from solar, wind, and hydrogen, as well as their peak power capacities. Long-term energy supply is further ensured by the direct connection between hydrogen generation and battery storage capacity. The models demonstrate that the hybrid system functions at high efficiency by minimizing losses during energy conversion and storage.

The potential of hybrid systems to offer dependable electricity and smooth out fluctuations in renewable energy supplies makes them vital for the future energy environment. Through the integration of modern storage technologies and complementing energy sources, these systems provide a solution to the problem of intermittency. This study concludes by demonstrating how hybrid renewable systems can be used to sustainably supply the world's energy needs. Combining battery storage with hydrogen fuel cells, solar energy, and wind power offers a reliable and expandable answer to the world's energy problems. Hybrid systems, such as the one this study suggests, will be essential to guaranteeing environmental sustainability and energy security as countries strive toward decarbonization and renewable energy targets.

REFERENCES

Abu-Rayash, A. and Dincer, I. 2021. Development and analysis of an integrated solar energy system for smart cities. *Sustainable Energy Technologies and Assessments*, *46*, p. 101170.

Abu-Rayash, A. and Dincer, I. 2023. Development and assessment of an integrated wind-solar based energy system for sustainable communities. *Energy Conversion and Management*, *277*, p. 116680.

Almutairi, A., Albagami, N., Almesned, S., Alrumayh, O. and Malik, H. 2023. Electric vehicle load estimation at home and workplace in Saudi Arabia for grid planners and policy makers. *Sustainability*, 15(22), 15878. http://dx.doi.org/10.3390/su152215878.

Amoussou, I., Paddy, E. Y., Agajie, T. F., Ibrahim, F. S., Agajie, E. F., Nsanyuy, W. B., Bajaj, M., & Mohammadi, S. A. D. (2024). Enhancing residential energy access with optimized stand-alone hybrid solar-diesel-battery systems in Buea, Cameroon. *Scientific Reports*, 14(1), p. 15543. https://doi.org/10.1038/s41598-024-66582-0.

Badri, Y. I., Sudabattula, S. K., Hussain, I., Malik, H. and Márquez, F. P. G. 2023. Ultracapacitor as selectable energy buffer in electric vehicle application. *Journal of Energy Storage*, 73, 109200. http://dx.doi.org/10.1016/j.est.2023.109200.

Badwawi, R. A., Abusara, M. and Mallick, T. 2015. A review of hybrid solar PV and wind energy system. *Smart Science*, *3*(3), pp. 127–138.

Bansal, P., Kumar, R., Kumar, A. and Dasig, D. Jr. 2024. *Artificial Intelligence and Communication Techniques in Industry 5.0*. Boca Raton, FL: CRC Press. https://doi.org/10.1201/9781003494027.

Bekele, G. and Palm, B. 2010. Feasibility study for a standalone solar–wind-based hybrid energy system for application in Ethiopia. *Applied Energy*, *87*(2), pp. 487–495.

Bhandari, P., Padalia, D., Ranakoti, L., Khargotra, R., András, K. and Singh, T. 2023. Thermo-hydraulic investigation of open micro prism pin fin heat sink having varying prism sides. *Alexandria Engineering Journal*, 69, 457–468. https://doi.org/10.1016/j.aej.2023.02.016.

Bhatt, A., Ongsakul, W. and Pawar, J. 2024. Optimal energy management system for carbon–neutral microgrid integrating second-life batteries and crypto mining devices. *Sustainable Energy Technologies and Assessments*, *64*, p. 103686. 10.1016/j.seta.2024.103686.

Chakrabarty, D., Patil, P. and Sharma, K. 2017. Experimental study on fluid flow and heat transfer from a rectangular prism approaching the wall of a wind tunnel. *Heat Transfer—Asian Research*, 46(7), 793–814. https://doi.org/10.1002/htj.21244.

Chávez-Ramírez, A. U., Vallejo-Becerra, V., Cruz, J. C., Ornelas, R., Orozco, G., Munoz-Guerrero, R. and Arriaga, L. G., 2013. A hybrid power plant (solar–wind–hydrogen) model based in artificial intelligence for a remote-housing application in Mexico. *International Journal of Hydrogen Energy*, *38*(6), pp. 2641–2655.

Couto, A. and Estanqueiro, A. 2023. Wind power plants hybridised with solar power: A generation forecast perspective. *Journal of Cleaner Production*, *423*, p. 138793.

Das, S., Thakur, P., Singh, A. K. and Singh, S. N., 2022. Optimal management of vehicle-to-grid and grid-to-vehicle strategies for load profile improvement in distribution system. *Journal of Energy Storage*, *49*, p. 104068. http://dx.doi.org/10.1016/j.est.2022.104068.

Ding, Z., Hou, H., Yu, G., Hu, E., Duan, L. and Zhao, J., 2019. Performance analysis of a wind-solar hybrid power generation system. *Energy Conversion and Management*, *181*, pp. 223–234.

Goyal, S., Mishra, S. and Bhatia, A., 2017. Modeling of solar/wind hybrid energy system usingMTALAB simulink. *International Journal of Pharmacy Research & Technology (IJPRT)*, 7(2), pp.30–33.

Gu, W., Ma, T., Ahmed, S., Zhang, Y. and Peng, J. 2020. A comprehensive review and outlook of bifacial photovoltaic (bPV) technology. *Energy Conversion and Management*, *223*, p.113283.

Gupta, S. C., Kumar, Y. and Agnihotri, G. 2007. Optimal sizing of solar-wind hybrid system. In *IET-UK International Conference on Information and Communication Technology in Electrical Sciences (ICTES 2007)* (pp. 282–287). Stevenage, UK: IET.

Hirose, T. and Matsuo, H. 2011. Standalone hybrid wind-solar power generation system applying dump power control without dump load. *IEEE Transactions on Industrial Electronics*, *59*(2), pp. 988–997.

Jamil, M., He, N., Wei, Z., Gupta, M. K. and Khan, A. M. 2023. A novel quantifiable approach of estimating energy consumption, carbon emissions and cost factors in manufacturing of bearing steel based on triple bottom-line approach. *Sustainable Materials and Technologies*, 36, e00593. http://dx.doi.org/10.1016/j.susmat.2023.e00593.

Jenkins, P., Elmnifi, M., Younis, A. and Emhamed, A. 2019. Hybrid power generation by using solar and wind energy: Case study. *World Journal of Mechanics*, *9*(4), pp. 81–93.

Kabalci, E. 2013. Design and analysis of a hybrid renewable energy plant with solar and wind power. *Energy Conversion and Management*, *72*, pp. 51–59.

Kanouni, B., Badoud, A. E., Mekhilef, S., Bajaj, M. and Zaitsev, I. 2024. Advanced efficient energy management strategy based on state machine control for multi-sources PV-PEMFC-batteries system. *Scientific Reports*, 14(1), 7996. https://doi.org/10.1038/s41598-024-58785-2.

Khare, V., Nema, S. and Baredar, P. 2016. Solar–wind hybrid renewable energy system: A review. *Renewable and Sustainable Energy Reviews*, *58*, pp. 23–33.

Kumar, A., Kumar, A. and Kumar, A. 2024. *Performance Characterization of Lubricants*. Boca Raton, FL: CRC Press. https://doi.org/10.1201/9781003535447.

Kumar, A., Singh, V. P., Meena, C. S. and Dutt, N. 2023. *Thermal Energy Systems: Design, Computational Techniques and Applications*. Boca Raton, FL: CRC Press. https://doi.org/10.1201/9781003395768.

Liu, L., Wang, Z., Wang, Y., Wang, J., Chang, R., He, G., Tang, W., Gao, Z., Li, J., Liu, C. and Zhao, L., 2020. Optimizing wind/solar combinations at finer scales to mitigate renewable energy variability in China. *Renewable and Sustainable Energy Reviews*, *132*, p. 110151.

Liu, T., Li, J., Yang, Z. and Duan, Y. 2024. Evaluation of the short-and long-duration energy storage requirements in solar-wind hybrid systems. *Energy Conversion and Management*, *314*, p. 118635.

Meena, C. S., Kumar, A., Singh, V. P. and Ghosh, A. 2024. *Sustainable Technologies for Energy Efficient Buildings*. Boca Raton, FL: CRC Press. https://doi.org/10.1201/9781003496656.

Molu, R. J. J., Tripathi, B., Mbasso, W. F., Naoussi, S. R. D., Bajaj, M., Wira, P. and Misak, S. 2024. Advancing short-term solar irradiance forecasting accuracy through a hybrid deep learning approach with Bayesian optimization. *Results in Engineering*, *23*, p. 102461. https://doi.org/10.1016/j.rineng.2024.102461.

Nasser, M., Megahed, T. F., Ookawara, S. and Hassan, H. 2022. Performance evaluation of PV panels/wind turbines hybrid system for green hydrogen generation and storage: Energy, exergy, economic, and enviroeconomic. *Energy Conversion and Management*, *267*, p.115870.

Punyavathi, R., Pandian, A., Singh, A. R., Bajaj, M., Tuka, M. B. and Blazek, V. 2024. Sustainable power management in light electric vehicles with hybrid energy storage and machine learning control. *Scientific Reports*, *14*(1), p. 5661. https://doi.org/10.1038/s41598-024-55988-5.

Rehman, S., Alam, M. M., Alhems, L. M. and Rafique, M. M. 2018. Horizontal axis wind turbine blade design methodologies for efficiency enhancement—A review. *Energies*, *11*(3), p. 506.

Rekioua, D., Mokrani, Z., Kakouche, K., Rekioua, T., Oubelaid, A., Logerais, P. O., . . . Ghoneim, S. S. 2023. Optimization and intelligent power management control for an autonomous hybrid wind turbine photovoltaic diesel generator with batteries. *Scientific Reports*, 13(1), 21830. http://dx.doi.org/10.1038/s41598-023-49067-4.

Sahri, Y., Tamalouzt, S., Belaid, S. L., Bajaj, M., Ghoneim, S. S., Zawbaa, H. M. and Kamel, S. 2023. Performance improvement of hybrid system based DFIG-Wind/PV/Batteries connected to DC and AC grid by applying intelligent control. *Energy Reports*, 9, 2027–2043. https://doi.org/10.1016/j.egyr.2023.01.021.

Shankar, A. 2023. Efficient data interpretation and artificial intelligence enabled IoT based smart sensing system. *Artificial Intelligence Review*, 56(12), 15053–15077. https://doi.org/10.1007/s10462-023-10519-y.

Singh, A. R., Kumar, R. S., Bajaj, M., Khadse, C. B. and Zaitsev, I. 2024. Machine learning-based energy management and power forecasting in grid-connected microgrids with multiple distributed energy sources. *Scientific Reports*, 14(1), 19207. https://doi.org/10.1038/s41598-024-70336-3.

Temiz, M. and Dincer, I. 2022. Development of solar and wind based hydrogen energy systems for sustainable communities. *Energy Conversion and Management*, *269*, p. 116090.

Temiz, M. and Javani, N. 2020. Design and analysis of a combined floating photovoltaic system for electricity and hydrogen production. *International Journal of Hydrogen Energy*, *45*(5), pp. 3457–3469.

Tiwari, S., Swaminathan, M., Harender and Singh, D. B. 2022. Performance enhancement of the photovoltaic system with different cooling methods. *Environmental Science and Pollution Research*, 29(30), 45107–45130. https://link.springer.com/article/10.1007/s11356-022-20330-x.

Verma, T., Singh, T. S., Rajak, U., Nashine, P., Dwivedi, G. and Kumar, A. 2024. *Clean Energy: Technology, Advances and Applications*. Boca Raton, FL: CRC Press. https://doi.org/10.1201/9781003521341.

Yadav, D., Awasthi, M. K. and Kumar, A. 2025. *Next Generation Renewable Thermal Energy Harvesting, Conversion & Storage Technologies*. Elsevier.

Yang, H., Lu, L. and Zhou, W. 2007. A novel optimization sizing model for hybrid solar-wind power generation system. *Solar Energy*, *81*(1), pp. 76–84.

Yang, H., Wei, Z. and Chengzhi, L. 2009. Optimal design and techno-economic analysis of a hybrid solar–wind power generation system. *Applied Energy*, *86*(2), pp. 163–169.

Index

FSC
www.fsc.org
MIX
Papier | Fördert gute Waldnutzung
FSC® C083411

Kolibri 360 GmbH
Ferdinand-Jühlke-Straße 7
99095 Erfurt, Deutschland
produktsicherheit@kolibri360.de